Biomedical Library

Queen's University Belfast

Tel: 028 9097 2710

E-mail: biomed.info@qub.ac.uk

For due dates and renewals:

QUB borrowers see 'MY ACCOUNT' at

http://library.qub.ac.uk/qcat

or go to the Library Home Page

HPSS borrowers see 'MY ACCOUNT' at

www.honni.qub.ac.uk/qcat

This book must be returned not later than its due
date, but is subject to recall if in demand

Fines are imposed on overdue books

The Practical Approach Series

SERIES EDITOR

B. D. HAMES
Department of Biochemistry and Molecular Biology
University of Leeds, Leeds LS2 9JT, UK

★ indicates new and forthcoming titles

Affinity Chromatography
★ Affinity Separations
Anaerobic Microbiology
Animal Cell Culture
  (2nd edition)
Animal Virus Pathogenesis
Antibodies I and II
★ Antibody Engineering
★ Antisense Technologies
★ Applied Microbial
  Physiology
Basic Cell Culture
Behavioural Neuroscience
Bioenergetics
Biological Data Analysis
Biomechanics—Materials
Biomechanics—Structures and
  Systems
Biosensors
Carbohydrate Analysis
  (2nd edition)
Cell–Cell Interactions
The Cell Cycle
Cell Growth and Apoptosis
Cellular Calcium

Cellular Interactions in
  Development
Cellular Neurobiology
★ Chromatin
Clinical Immunology
★ Complement
Crystallization of Nucleic
  Acids and Proteins
Cytokines (2nd edition)
The Cytoskeleton
Diagnostic Molecular Pathology
  I and II
★ DNA and Protein Sequence
  Analysis
DNA Cloning 1: Core
  Techniques (2nd edition)
DNA Cloning 2: Expression
  Systems (2nd edition)
★ DNA Cloning 3: Complex
  Genomes (2nd edition)
★ DNA Cloning 4: Mammalian
  Systems (2nd edition)
Electron Microscopy in
  Biology
Electron Microscopy in
  Molecular Biology

Mutation Detection
A Practical Approach

Edited by

R. G. H. COTTON
Mutation Research Centre
Victoria, Australia

E. EDKINS
Joint Women's and Children's Hospital
Perth, Australia

and

S. FORREST
Murdoch Institute
Victoria, Australia

OXFORD UNIVERSITY PRESS
Oxford New York Tokyo

Oxford University Press, Great Clarendon Street, Oxford OX2 6DP

Oxford   New York
Athens Auckland Bangkok Bogota Bombay Buenos Aires
Calcutta Cape Town Dar es Salaam Delhi Florence Hong Kong
Istanbul Karachi Kuala Lumpur Madras Madrid Melbourne
Mexico City Nairobi Paris Singapore Taipei Tokyo Toronto Warsaw
and associated companies in
Berlin Ibadan

Oxford is a trade mark of Oxford University Press

Published in the United States
by Oxford University Press Inc., New York

A catalogue record for this book is available from the British Library

Library of Congress Cataloging in Publication Data
(Data available)
ISBN 0 19 963657 5 (Hbk)
ISBN 0 19 963656 7 (Pbk)

Typeset by Footnote Graphics, Warminster, Wilts
Printed in Great Britain by Information Press, Ltd, Eynsham, Oxon.

Contents

Contents

7. Detection of mutations by hybridization with sequence-specific oligonucleotide probes   113

Randall K. Saiki and Henry A. Erlich

Contributors

ASFHIN AHMADIAN
Department of Biochemistry and Biotechnology, Kungliga Tekniska Högskolan, Royal Institute of Technology, Teknikringen 34, S-100 44 Stockholm, Sweden

GISELA BARBANY
Department of Medical Genetics, Box 589 BMC, Se-751 23 Uppsala, Sweden.

CHERIF BELDJORD
Laboratory of Biochemistry and Molecular Genetics, CHU, Cochin, 123 Blvd Pord Royal, 75014 Paris Cedex, France.

CYNTHIA D. K. BOTTEMA
Deparment of Animal Science, Waite Campus, University of Adelaide, Glen Osmond, South Australia 5064, Australia

RICHARD G. H. COTTON
Mutation Research Centre, Daly Wing, St Vincent's Hospital, 41 Victoria Parade, Fitzroy 3065, Melbourne, Victoria, Australia.

MICHAEL DEAN
National Cancer Institute, Frederick Cancer Research and Development Center, Frederick, MD 21702, USA.

JOHAN T. DEN DUNNEN
MGC, Department of Human Genetics, Leiden University, Wassenaarseweg 72, 2333 AL Leiden, The Netherlands.

EDWARD EDKINS
Joint Women's and Children's Hospital, Perth, Australia.

HENRY A. ERLICH
Department of Human Genetics, Roche Molecular Systems, 1145 Atlantic Avenue, Alameda, CA, 94501 USA.

ANNE ESTREICHER
Swiss Institute for Experimental Cancer Research (ISREC), 1066 Epalinges, Switzerland.

JEAN-MICHEL FLAMAN
Laboratory of Molecular Genetics, CHU de Rouen, 76031 Rouen, France.

RICCARDO FODDE
MGC, Department of Human Genetics, Leiden University, Wassenaarseweg, 72, 2333 AL Leiden, The Netherlands.

Contributors

SUE FORREST
Murdoch Institute, Royal Children's Hospital, Flemington Road, Parkville, Victoria, Australia.

THIERRY FREBOURG
Laboratory of Molecular Genetics, CHU de Rouen, 76031 Rouen, France.

BERNARD GERRARD
SAIC-Frederick, Frederick Cancer Research and Development Center, Frederick, MD 21702, USA.

FRANCESCO GIANNELLI
Division of Medical and Molecular Genetics, United Medical and Dental School of Guy's and St Thomas' Hospitals, Guy's Campus, Guy's Tower, London, SE1 9RT, UK.

PETER M. GREEN
Division of Medical and Molecular Genetics, United Medical and Dental School of Guy's and St Thomas' Hospitals, Guy's Campus, Guy's Tower, London, SE1 9RT, UK.

ANETTE HAGBERG
Department of Medical Genetics, Box 589 BMC, Se-751 23 Uppsala, Sweden.

KENSHI HAYASHI
Division of Genome Analysis, Institute of Genetics Information, Kyushu University, Maidashi 3-1-1, Higashi-ku, Fukuoka 812-82, Japan.

RICHARD IGGO
Swiss Institute for Experimental Cancer Research (ISREC), 1066 Epalinges, Switzerland.

MASAKAZU INAZUKA
Division of Genome Analysis, Institute of Genetics Information, Kyushu University, Maidashi 3-1-1, Higashi-ku, Fukuoka 812-82, Japan.

YOUJI KUKITA
Division of Genome Analysis, Institute of Genetics Information, Kyushu University, Maidashi 3-1-1, Higashi-ku, Fukuoka 812-82, Japan.

MAREK KWIATKOWSKI
Department of Medical Genetics, Box 589 BMC, Se-751 23 Uppsala, Sweden.

ULF LANDEGREN
Department of Medical Genetics, Box 589 BMC, Se-751 23 Uppsala, Sweden.

L. S. LERMAN
Department of Biology, Massachusetts Institute of Technology, Cambridge, MA 02139, USA.

Contributors

JOAKIM LUNDEBERG
Department of Biochemistry and Biotechnology, Kungliga Tekniska Högskolan, Royal Institute of Technology, Teknikringen 34, S-100 44 Stockholm, Sweden

HIROKI NAGASE
Beatson Institute for Cancer Research, Garscube Estate, Switchback Road, Bearsdon, Glasgow G61 1BD, UK.
(present) ONYX Pharmaceuticals, 3031 Research Drive, Richmond, CA 94806, USA.

YUSUKE NAKAMURA
Human Genome Centre, Institute of Medical Science, The University of Tokyo, 4-6-1 Shirokanedai, Minato-ku, Tokyo 108, Japan.

MATS NILSSON
Department of Medical Genetics, Box 589 BMC, Se-751 23 Uppsala, Sweden.

JÜRI PARIK
Eesti Biokeskus, Tartu University, Tähetom Toomei, 202400 Tartu, Estonia.

SUSAN J. RAMUS
CRC Human Cancer Genetics Research Group, Level 3, Laboratory Block, Box 238, Addenbrooke's Hospital, Cambridge, CB2 2QQ, UK.

RANDALL K. SAIKI
Department of Human Genetics, Roche Molecular Systems, 1145 Atlantic Avenue, Alameda, CA, 94501, USA.

MARTINA SAMIOTAKI
Hellenic Pasteur Institute, 127 Vas. Sofias Avenue, Athens 11 521, Greece.

STEVE S. SOMMER
Beckman Research Institute, City of Hope, 1450 east Duarte Road, Duarte, CA 91010-0269, USA.

ANN-CHRISTINE SYVÄNEN
Department of Human Molecular Genetics, National Public Health Institute, Mannerheimintie, 166, 00300, Helsinki, Finland.

TOMOKO TAHIRA
Division of Genome Analysis, Institute of Genetics Information, Kyushu University, Maidashi 3-1-1, Higashi-ku, Fukuoka 812-82, Japan.

MATHIAS UHLÉN
Department of Biochemistry and Biotechnology, Kungliga Tekniska Högskolan, Royal Institute of Technology, Teknikringen 34, S-100 44 Stockholm, Sweden

ROB B. VAN DER LUIJT
MGC, Department of Human Genetics, Leiden University, Wassenaarseweg 72, 2333 AL Leiden, The Netherlands.

Contributors

CECILIA WILLIAMS
Department of Biochemistry and Biotechnology, Kungliga Tekniska Högskolan, Royal Institute of Technology, Teknikringen 34, S-100 44 Stockholm, Sweden

RIMA YOUIL
Merck & Co. Inc., Merck Research Laboratories, Department of Human Genetics, WP26A-3000, Sumneytown Pike, West Point, PA 19486, USA.

Abbreviations

APC	adenomatous polyposis coli gene
APS	ammonium persulfate
ASA	allele-specific amplification
ASO	allele-specific olignucleotides
BMD	Becker muscular dystrophy
C	Cross-linking (% acrylamide to *bis*-acrylamide)
CCM	chemical cleavage of mismatch
CDCE	constant denaturant capillary electrophoresis
CDGE	constant denaturant gel electrophoresis
cDNA	complementary DNA
CE	capillary electrophoresis
CE-SSCP	capillary electrophoresis SSCP
CF	cystic fibrosis
dCTP	deoxycytidine triphosphate
DEPC	diethylpyrocarbonate
DGGE	denaturing gradient gel electrophoresis
DMD	Duchenne muscular dystrophy
DMSO	dimethyl sulphoxide
DNA	deoxyribonucleic acid
dNTP	nucleoside triphosphates
EDTA	ethylenediaminetetraacetic acid
EMC	enzyme mismatch cleavage
EMD	enzymatic mutation detection
EMD™	Enzymatic Mutation Detection™
FAP	familial adenomatous polyposis
HA	heteroduplex analysis
HPLC	high pressure liquid chromatography
MDE	mutation detection enhancing
MDE™	Mutation Detection Enhanced™
Mes	2 (N-morpholino)-ethane sulphonic acid
MFS	Marfan syndrome
mRNA	messenger RNA
NIRCA	non-isotopic RNase cleavage assay
OLA	oligonucleotide ligation assay
PAMSA	PCR amplification of multiple specific alleles
PASA	PCR amplification of specific alleles
PBS	phosphate buffered saline
PCR	polymerase chain reaction
PKU	phenylketonuria
PNK	polynucleotide kinase

PTT	protein truncation test
REF	restriction endonuclease fingerprinting
RNA	ribonucleic acid
RNase	ribonuclease
RPA	ribonuclease protection assay
RT	reverse transcriptase
SDS	sodium dodecyl sulphate
SSCP	single-stranded conformation polymorphism
SSO	sequence specific oligonucleotide
SSOP	sequence-specific oligonucleotide probes
TBE	Tris, borate, EDTA
TDGGE	thermal denaturing gradient gel elution
TdT	terminal deoxyribonucleotidyl transferase
TE	Tris, EDTA
TEMED	tetramethylethylenediamine
TGGE	thermal denaturing gradient electrophoresis
T_m	melting temperature
TMB	tetramethylbenzidine
TME	Tris-Mes-EDTA
TSGE	temperature sweep gel electrophoresis
TTGE	temporal temperature gradient electrophoresis
UV	ultraviolet
VNTR	variable number of tandem repeats

Introduction

SUE FORREST

Mutation detection is becoming vital to the whole of biology, let alone medicine. In medical research it is fundamental to disease gene isolation, mutation spectrum studies, and diagnosis. It is most important in the area of cancer and cancer research where increasing numbers of cancer genes have been isolated over recent years. In biology, commercially important genes can be identified by the mutations they contain. Strains of organisms can be characterized because they differ by a number of mutations. It is not surprising then that there is a rising interest in mutations and mutation detection.

Mutation detection is time-consuming and expensive. This partly explains the range of methods now available to detect mutations as there is no perfect method of mutation detection yet described. Two classes of mutation detection techniques can be envisaged. The first is known as a scanning mode which searches for unknown mutations. The second is a diagnostic mode which tests for known mutations. The scanning methods covered in this volume were chosen mainly because of their current usage, their simplicity, their ability to detect near 100% of mutations, and the ability of some to scan longer lengths. The methods for known mutation detection comprise the four best known current methods. The TAQMAN™ method for mutation detection was specifically not covered because a kit has been available for some time and few instructions are necessary beyond those given by the manufacturer.

In the case of scanning methods, there are simple methods that do not detect all mutations ranging to more complex methods which detect close to 100% of mutations. The mutation detector will find it hard to choose amongst the methods, but often factors such as budget, laboratory and personnel experience, per cent detection required, and the project itself will be taken into account. With methods for detecting known mutations, similar considerations are present. Some methods are more familiar, such as allele-specific amplification (ASA) and allele-specific oligonucleotides (ASO), and may be chosen for that reason. Others may be chosen because of their ease of application to large samples.

In all the techniques, there tends to be a greater uptake if a kit is available and this is possible to varying degrees for all the methods ranging from special gels for the physical methods to enzyme preparations. Enzyme cleavage of mismatch (EMC) is clearly heading in this direction and there is a 'kit' for ribonuclease.

To assist with choice of methodology, mutation detection in the field of human genetic diseases will be used as an example. Assuming the gene has been cloned, direct testing for the mutations in the gene is possible. The gene will be tested for already described mutations or searched for new mutations. The types of alterations that such tests need to be able to detect include deletions or insertions, single base pair substitutions, or trinucleotide repeat expansions.

Before discussing the different techniques in greater detail, there are a number of limitations associated with direct gene testing which need to be understood when designing a mutation detection strategy. First, the number of different mutations in a gene will influence the type of method used. In sickle cell anaemia for example, one mutation in the β-globin gene is responsible for the disorder. In contrast, over 600 different mutations have so far been identified in the cystic fibrosis transmembrane conductance regulator gene. Therefore different strategies will need to be considered.

The size of the gene and the number of exons is also important. For both β-thalassaemia and haemophilia A, over 100 different mutations have been identified in the β-globin gene and factor VIII gene respectively. However, the β-globin gene is only 1.5 kb long, has a 650 bp mRNA, and three exons, whilst the FVIII gene is greater than 186 kb long, has a 9 kb mRNA, and 26 exons. Even though the same number of mutations needs to be searched for, the scale of the problem differs quite dramatically between the two genes. The choice of which mutation detection method to use is highly gene-dependent.

Instead of testing each exon of a gene individually for mutations, attention has turned towards using mRNA as the source material for mutation detection which allows more of the gene to be scanned in one assay. In addition, it offers the benefit of being able to detect RNA splicing defects which may be missed in a genomic DNA analysis. However, not all genes are expressed in readily accessible tissues. For example, if one was to look for mutations in a liver-specific gene, then a liver biopsy would be required to isolate mRNA. This is not a practical possibility for most patients.

Therefore the potential to study so-called 'ectopic' or 'illegitimate' transcripts of genes presented an exciting opportunity. The existence of very low levels of perfectly spliced mRNA was identified in cells that were not expected to express the mRNA. This was first shown by Chelly *et al.* (1) who studied the expression of the dystrophin gene in non-muscle tissue such as fibroblasts and lymphocytes. The exact mechanism for the presence of these transcripts has not been elucidated. Sarkar and Sommer (2) confirmed the existence of illegitimate transcripts when examining the possibility of accessing a number of different mRNAs from lymphocytes. They showed, for example, that liver-specific phenylalanine hydroxylase and factor VIII mRNA could be amplified. This type of analysis usually requires two rounds of PCR amplification, using a second set of PCR primers nested within the first to obtain enough product to analyse. Further studies have shown that this approach is robust and has been used to detect mutations in patients with a variety of different genetic diseases (for reviews, see refs 3 and 4).

2

However, there are a number of pitfalls that could possibly occur and caution must be exercised in interpretation of the results. First, aberrant splicing of the gene may occur in the tissues used to access the illegitimate transcripts. Therefore the structure of the illegitimate transcripts may not always faithfully reflect the structure of the transcript in the normal expressing cells. These products may be confused for the mutation and it is important to include normal samples from the same cell type as the patients to differentiate such findings. For example, in the phenylalanine hydroxylase gene, exon 11 is preferentially spliced out in lymphocytes (Susan Ramus, personal communication). Given that some of these alternatively spliced forms may exist, correctly spliced illegitimate transcripts are also present in all studies so far performed and thus serve as bona fide sources of mRNA for mutation analysis.

Another observation is that different alleles of a gene may be expressed in differing amounts. This has been studied in detail by a number of researchers. In particular, a series of patients who were compound heterozygotes for two known mutations were studied and it was found that from different aliquots of the same cDNA preparation, one or other allele solely was represented following two rounds of PCR (Susan Ramus, personal communication). This was concluded to be due to such a low level of the mRNA present in the initial sample for this particular gene and tissue. The recommendation was therefore to mix the products from two independent PCR reactions to attempt to ensure that equal amounts of the two alleles are present.

Another concern with using mRNA as the source material for mutation detection is the effect of certain mutations on the stability of the mRNA. In particular, it has been shown that even nonsense codons can affect both the splicing of exons in which they are contained as well as decreasing the stability of the mRNA. The exact mechanism for such a finding is yet to be elucidated. To counter this concern, some laboratories analyse genomic DNA and mRNA routinely.

Many of the diagnostic techniques focus on the detection of single base pair substitutions though many could also be applied to the detection of small deletions and insertions. The allele-specific oligonucleotide (ASO) method involves designing two oligonucleotides, one perfectly matching the mutant and one the normal sequence. The normal oligonucleotide will not bind to the mutant and vice versa. The oligos are hybridized to the samples following a PCR reaction. Allele-specific amplification (ASA) means that amplification only occurs if the primer perfectly matches the target sequence. Two primers are designed such that one is complementary to the normal and one to the mutant sequence.

In primer extension or minisequencing, a primer is designed 5′ to the base that varies in the mutated allele and can only be extended when the base complementary to the one present in the allele is provided in the reaction mix. Ligation assay involves the design of three oligos, two that differ by the final base and one common oligo. If the test oligo hybridizes perfectly to the sequence, the two oligos can be joined together by ligase. If there is a mismatch at the 3′ end of the 5′ primer, no ligation occurs. Creation or abolition of a restriction site is widely used for known mutation detection. Also, a restriction site can be engineered into the PCR product to facilitate discrimination of alleles (AIRS) if one is not created or abolished by the mutation. This is simple, is already in widespread use, and is described in ref. 5.

The other group of methodologies to consider are the scanning methodologies. The first group of methods was based on the aberrant migration of mutant molecules during electrophoresis. These include:

(a) Denaturing gradient gel electrophoresis, where either a chemical or temperature gradient is used.

(b) Heteroduplex analysis in a non-denaturing gel, where a heteroduplex with a mismatch has a different mobility from the corresponding homoduplex.

(c) Single-stranded conformational polymorphism analysis, where single-stranded molecules with mutations migrate differently.

These methods, whilst simple, share the disadvantage that the position of the mismatch is not pin-pointed and it remains questionable if all mutations can be detected.

The second group of methods relies on cleavage of RNA or DNA molecules prior to analysis. These include ribonuclease cleavage, chemical, and more recently, T4 endonuclease VII cleavage. These techniques offer two significant advantages: the position of the mutation can be localized, and longer fragments of DNA (greater than 1 kb) can be tested.

Sequencing as a method for detecting unknown mutations (as distinct from defining them), should not be omitted even though many of the scanning methods aim to eliminate this step for defining if a mutation is present. This volume describes the single colour system rather than the more widely known four colour system.

The problem which remains with many of the scanning methodologies is 'does this mutation actually cause disease'? This is where the functional assays come into their own right and for some genes, they are now considered as the first mutation scanning test to be performed. Functional methods are highly desirable, as it can tell the investigator whether there is a *deleterious* mutation present which the other scanning methods are unable to achieve. There are two main types of functional tests currently available which are referred to as protein truncation testing (or *in vitro* synthesized protein assay), and complementation assays. Of course for many years expression of genes with putative mutations and assays of enzyme activity have existed and are still vital.

In familial polyposis colorectal cancer, greater than 85% of mutations in the gene result in premature termination of protein synthesis. Therefore a test was developed that could detect such alterations—the protein truncation test. The starting material can either be genomic DNA or RNA that has been copied into cDNA. A 5′ primer is designed that incorporates a T7 RNA polymerase binding site, a ribosome binding site, and the eucaryotic transcription initiation signal. The starting AUG codon must be in-frame with the reading frame of the test protein. Following PCR, each product has the extension on the 5' end. The product is then added to an *in vitro* transcription/translation system where RNA then protein is produced. Patients that have a mutation in their DNA which results in the introduction of a premature stop codon in the mRNA will produce a shortened protein product.

The list of disorders which are applicable to PTT testing is growing and in particular includes many of the cancer-causing genes. These include FAP, hereditary non-polyposis colon cancer (at least two of the four genes responsible), neurofibromatosis types 1 and 2, and mutations in BrcaI that can cause breast cancer.

The second functional assay is complementation. An example of this is testing for mutations that affect the function of p53. The assay starts with mRNA from which a PCR product of the region of interest is generated. The middle portion of p53 represents the DNA binding domain. This can be cloned into a specially designed gap vector using homologous recombination in yeast. An ade2 reporter plasmid is used in a yeast that is deficient in *ade2*. If wt p53 is produced, it can bind to the p53 DNA binding sites and activate the transcription of the *ade2* gene, overcoming the metabolic block, resulting in the production of white colonies (as the yeast lacks the ability to synthesize adenine itself). If the p53 has lost its DNA binding function, no adenine is produced and a red colour results from a build up of a coloured intermediate. The assay also allows both alleles to be cloned independently.

Functional methods for many other genes are being sought besides the examples given here but have proven difficult to establish. Some protein quality information can be given by the PTT test which specifically detects truncated and thus, faulty protein, but in its current format, PTT will not detect faulty protein due to charge changes.

Solid phase technology for mutation detection, both in scanning and diagnostic modes, has made manipulations easier, provides the opportunity for automation, and may offer the possibility to avoid electrophoresis. The initial usage was membrane supports for oligo hybridization assays where either the target or the oligo probe was bound to the support. Manifold supports resemble gel combs and molecules of interest are bound to the surface of a multipronged support and subjected to a series of reactions culminating in the final placement of the molecules on the comb into an analytical system. The binding capacity of the surface is increased 800-fold by fusing porous particles to polystyrene surfaces. These supports have been used for ligase-mediated gene diagnostics, DNA sequencing, or scanning using T4 endonuclease VII. Magnetic beads have also been used for DNA sequencing.

Whilst not dealt with in this volume as it is not widespread, there is much expectation that the chip assay for known and unknown mutations will one day become a reality. Here overlapping sets of oligos, say eight bases long, are designed to cover the entire length of the gene to be assayed. Each position in the target sequence is queried by a set of four probes on the chip which are identical except for one position, the substitution position in the centre. This position is either A, C , T, or G. Application to genes are now appearing and the technique has been recently reviewed by Ginot (6).

The methods mentioned in this introduction are described in detail in the following chapters. Many of the methods could be used for the same research studies and it is now left to the researcher to make that final choice bearing all the previous considerations in mind.

References

1. Chelly, J., Kaplan, J-C., Maire, P., Gautron, S., and Kahn, A. (1988). *Nature*, **333**, 858.
2. Sarkar, G. and Sommer, S. S. (1989). *Science*, **244**, 331.
3. Cooper, D. N., Berg, L-P., Kakkar, V. V., and Reiss, J. (1994). *Ann. Med.*, **26**, 9.
4. Kaplan, J-C., Kahn, A., and Chelly, J. (1992). *Hum. Mutat.*, **1**, 357.
5. Cotton, R. G. H. (1997). *Mutation detection*. Oxford University Press, UK.
6. Ginot, F. (1997). *Hum. Mutat.*, **10**, 1–10.

Single-strand conformation polymorphism analysis

KENSHI HAYASHI, YOUJI KUKITA, MASAKAZU INAZUKA, and TOMOKO TAHIRA

1. Introduction

SSCP (single-strand conformation polymorphism) is widely used for mutation detection because of its simplicity and versatility. In this method, a segment to be searched for a mutation is amplified by PCR from genomic DNA or cDNA, denatured, and separated by electrophoresis in media having sieving properties without denaturant (1). Mutation usually causes a mobility shift in electrophoresis, and so, is detected. Conformational change of the single-stranded mutant DNA is believed to be the reason for the mobility shift.

Conformation (i.e. tertiary structure) of single-stranded DNA is determined by intramolecular interactions which can change depending on physical conditions, e.g. temperature and ionic environment. Accordingly, separation of mutant DNA in SSCP varies depending on the conditions of electrophoresis. Though empirical rules on good separation of sequence variants are emerging (3), whether a certain mutation can be detected in a given condition is not predictable. However, sensitivity of PCR-SSCP is generally believed to be high if the fragments are short (2, 4). Because there is no theoretical assurance that sensitivity is a 100%, non-appearance of a new band does not prove absence of mutation.

Mutations detectable by PCR-SSCP are not restricted to fixed positions in the PCR amplified DNA fragments. Other techniques in a similar category, i.e. so-called scanning methods, are DGGE (denaturing gradient gel electrophoresis) (5), heteroduplex (6), and mismatch cleavage (7, 8) analyses. Among them, SSCP is the simplest technique. The technique also has the advantages of detecting and isolating mutant fragments that constitute a minor fraction, and characterizing them by sequencing after reamplification (9).

SSCP has been used to search for disease-associated unknown mutations in, for example, possibly oncogenic genes in cancer tissues (10, 11), and in the candidate hereditary disease genes in carriers or patients of the disease (12), to prove that the genes are indeed responsible. The method is also a valid choice in clinical diagnosis to prove the presence or absence of detectable mutations, e.g. to determine whether or not a child carries a mutant allele which has been detected in the parental genome.

The first part of this chapter (Section 2) describes SSCP analysis in polyacrylamide gel electrophoresis using ^{32}P-labelled DNA. Procedures of the analysis in capillary electrophoresis using fluorescence labelled DNA is described next (Section 3).

2. PCR-SSCP using polyacrylamide slab gel

High percentage detection depends on high resolution electrophoresis, and so a commonly adopted strategy is to radioisotopically label the DNA and detect them by autoradiography after separation in thin and long native polyacrylamide gels. Labelling PCR products with ^{32}P can be done either by primer labelling or internal incorporation of ^{32}P-labelled nucleotides during the amplification reaction. The internal labelling procedure is shown below, because the procedure is simpler and uses less radioactivity.

It has been empirically known that using Tris–borate–EDTA–glycerol as a buffer system in the electrophoresis (see *Protocol 4*), mutations can be detected at a high sensitivity (13). An important recent improvement is the use of low pH gel in the electrophoresis. In a model system, 14 out of 16 mutant fragments carrying different single base substitutions in 600–800 bp length showed clearly identifiable mobility shifts in a single electrophoretic run (3). Thus, the recommended range of fragment size for SSCP analysis may now be extended to up to 800 bp or perhaps longer, while in a conventional system, it has been generally believed to be up to approximately 300 bp (*Figure 1*).

Although we believe low pH gel system is the first choice of electrophoresis especially for long PCR products, the system does have a drawback, i.e. we occasionally experienced broadening of bands for some sequence contexts, and even smearing in extreme cases, though such cases are infrequent. The tendency to broaden and smear is also encountered when electrophoresis is carried out in Tris–borate–EDTA–glycerol when it is run at a low temperature, e.g. 4°C. It seems that in the conditions where mutations are more sensitively detected, the mobility is also more sensitive to temperature, and uneven temperature distribution, e.g. the difference between the temperatures at the centre of the gel and at the contact to the glass plates, may be sufficient to cause significant change of mobility. The exact reason, however, remains unexplained.

Once a mutation is detected as the presence of bands with shifted mobility, DNA in the bands is recovered, re-amplified, and sequenced to identify the nucleotide change (9). Mutant alleles which constitute as low as a few per cent can be unambiguously detected and characterized by this method.

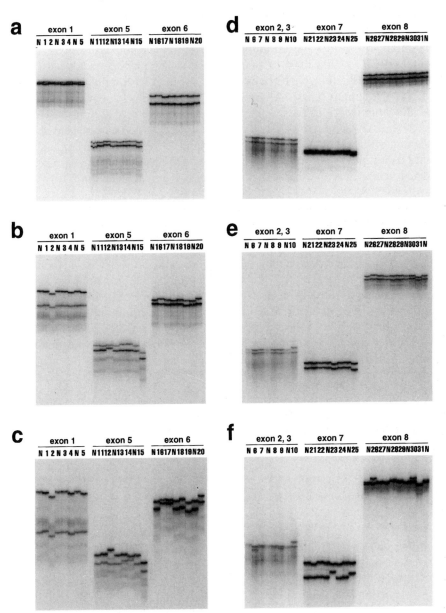

Figure 1. Shows the effects of pH on PCR-SSCP analysis. The samples were electrophoresed in 0.5 TBE (pH 8.3) for 2.8 hr (a) or 4.5 hr (d), in 0.5 TBE with 5% glycerol (pH 7.7) for 5.0 hr (b) or 7.5 hr (e), and in TPE (pH 6.8) for 5.0 hr (c) or 8.5 hr (f). Short PCR products of 424, 338, and 389 bps of Factor IX exons 1, 5, and 6, respectively (a, b, and c), and long PCR products of 629, 597, and 797 bps of Factor IX exons 2–3, 7 and 8, respectively (d, e, and f) were analyzed in separate gels. Lanes labeled N represents normal control. Samples of hemophilia patients were loaded in lanes labeled 1–31. (Adapted from ref. 3.)

2.1 PCR optimization and primer design

As in many other PCR-based mutation detection methods, it is important to optimize PCR so that little or no fortuitous fragments are produced in the amplification reaction. Sensitivity of PCR-SSCP gradually decreases by increasing fragment length. Therefore, it is preferable to set the primers so that the amplification does not exceed, e.g. 500 bp. Mutations in longer stretches of DNA sequence (such as the whole cDNAs) can be searched by designing primer sets so that the entire region is covered by overlapping short amplification units. Alternatively, long PCR products are cut by appropriate restriction enzymes and then examined by SSCP gel (see Section 2.10).

2.2 Pre-amplification and isolation by agarose gel electrophoresis

Starting from 'difficult' templates such as DNA of formalin-fixed and paraffin-embedded tissues or archaeological materials, success of PCR is stochastic. The low rate of success is likely to be attributable to a limited number of templates available in these samples. The same is true for PCR from dilute DNA solutions, e.g. few molecule PCR. In such situations, pre-amplification is an effective approach because the number of template molecules in PCR with $[\alpha\text{-}^{32}P]dCTP$ can be brought to a high and constant number, and unknown inhibitory activity possibly present in the original template solution need not be a problem. If the sample DNA is clean and plentiful, this step can be ignored.

Protocol 1.   Pre-amplification

Equipment and reagents

• Apparatus and reagents for agarose gel electrophoresis: 2% agarose mini-gel containing 0.1 µg/ml of ethidium bromide

• Spin filtrator: Ultrafree C3 GV filter (Millipore)

Method

1. Amplify target sequence in generally recommended conditions (using 100–200 µM of each of the four dNTPs and 1 µM of each of the two primers).

2. Separate the product by agarose gel electrophoresis.

3. Estimate the amount of DNA in the band of expected length from intensity of ethidium bromide staining.[a]

4. Cut out the bands of interest in approx. 2 × 5 × 6 mm plugs using a razor blade.

5. Place the plugs in the upper cups of a spin filtrator and centrifuge at 5000 *g* for 5 min.

---

6. Dilute the eluate in the lower tube, if necessary, to approx. 0.5 fmol/μl with 0.1 × TE,[b] and use as templates for labelling cycles of PCR following *Protocol 2*.

[a] Take 1 μl of the PCR product, mix with 5 μl of 5% glycerol, 0.05% bromophenol blue, and 0.05% xylene cyanol, and load onto gel. Also load a known amount of marker DNA, e.g. *Hae*III digested øX174 DNA, in the separate lane and run electrophoresis. The amount of sample DNA in the gel is estimated by referring the band intensity of samples to those of the marker DNA.
[b] The eluate in the lower tube (typically 30–40 μl) contains 40% (200 bp) to 20% (800 bp) of DNA in the plug (Kukita, unpublished data). 0.5 fmol of a 200 bp fragment is approx. 60 pg.

2.3 PCR using [^{32}P]deoxynucleotide triphosphate

In this protocol, the target sequence in genomic DNA, or pre-amplified and gel purified DNA (see Section 2.2) is amplified and labelled simultaneously by including [α-^{32}P]dCTP in the PCR. Primer concentrations and nucleotide concentrations in PCR with [α-^{32}P]dCTP are lower than generally recommended by suppliers of the enzymes or kits. This is to reduce consumption and possible hazard of radioactivity, and to ensure high specific activity of label in the PCR product. Low specific activity of the label may lead to overloading of samples to gel, resulting in reduced resolution. Overloading may also result in complication of the electrophoretogram because of an association of primers and single-stranded DNA in the sample solution before or during separation by electrophoresis (14).

Protocol 2. Amplification and ^{32}P-labelling

Reagents

- Primers: stored frozen at −20°C as 10 μM solution in H_2O
- [α-^{32}P]dCTP: 10 mCi/ml, 3000 Ci/mmol (Amersham)
- *Taq*/anti*Taq*: a 1:1 (v/v) mixture of *Taq* DNA polymerase (5 U/μl) (PE Applied Biosystems) and anti*Taq* antibody™ (1.1 mg/ml) (*Taq*Start Antibody, Clontech), prepared and stored as recommended by the supplier of the antibody[a]
- 25 mM $MgCl_2$

- 10 × PCR buffer: 0.5 M KCl, 0.1 M Tris–HCl pH 8.3
- 1.25 mM dNTP: a mixture of an equal concentration of dATP, dCTP, dGTP, and dTTP (Pharmacia), each at 1.25 mM
- Formamide dye: 95% deionized formamide, 20 mM Na_2EDTA pH 7.8, 0.05% bromophenol blue, 0.05% xylene cyanol—formamide (analytical grade, Merck) is deionized by the described procedures (15)

Method

1. Mix following solutions. This mixture is enough for ten reactions (total volume 40.5 μl).

- H_2O 25.0 μl
- 10 × PCR buffer   5.0 μl
- 25 mM $MgCl_2$ 5.0 μl
- Left primer   1.0 μl

Protocol 2.   *Continued*

- Right primer   1.0 μl
- 1.25 mM dNTP   1.0 μl
- [α-^{32}P]dCTP   2.0 μl
- *Taq*/anti*Taq*   0.5 μl

2. Divide the mixture into PCR tubes, 4 μl each.

3. Add 1 μl of template DNA (50 ng/μl of genomic DNA in 0.1 × TE, or 0.5 fmol/μl of pre-amplified and gel purified DNA, see *Protocol 1*), and vortex.

4. (This step is for users of thermal cycler without heated lid.) Overlay with mineral oil (15 μl/tube) and briefly spin to settle the layers.

5. Put tubes into thermal cycler and start cycling. Typically, 94 °C for 1 min, then, cycles of 94 °C for 30 sec and 60 °C or 65 °C for 2 min. 30–35 cycles are needed for genomic DNA. Ten cycles are sufficient for pre-amplified and gel purified DNA.

6. Follow *Protocol 3* to remove excess nucleotides. Stop the reaction at this stage by adding 45 μl of formamide dye, if the labelled PCR products are to be directly analysed by SSCP.

[a] Anti*Taq* antibody suppresses unwanted polymerase action during preparation of reaction mixture without template DNA in a batch, which is done at room temperature, and chain elongation can be initiated from primer:primer complexes if the polymerase is active (15). Recently, a package of *Taq* DNA polymerase with heat labile inhibitor (Ampli*Taq* GOLD, PE Applied Biosystems) has been made available. This heat activatable enzyme can also be used in place of *Taq*/anti*Taq*.

2.4 Removal of 3′ appendage

Sometimes, more than two bands per strand of one sequence context are observed if PCR products are directly applied and separated in SSCP gel. Appearance of the extraneous bands has been previously interpreted to be the presence of 'iso-conformers', i.e. more than two conformations for one sequence, in the SSCP conditions. However, most of the 'iso-conformers' in earlier observations in PCR-SSCP seem to be explained by 3′ heterogeneity of the PCR product (17; Kukita, unpublished observations).

Taq DNA polymerase has the characteristic of adding an extra nucleotide at each 3′ end of double-stranded DNA. This extra nucleotide, usually adenylate, does not base pair to the template strand. The percentage of ends having this appendage varies depending on the primer sequence and conditions of the reaction. Strands with appendage behave differently from those without in SSCP. Appendages can be effectively removed by treating with Klenow fragment of DNA polymerase I in the presence of nucleotides. The enzyme cleaves the appendage, but further hydrolysis by its 3′ exonuclease activity is prevented by the presence of the four nucleotides. This cleaning-up of the ends eliminates '*iso*-conformer' bands and simplifies the autoradiogram of SSCP.

Whether to remove the appendage or not is, however, optional. In many cases, the proportion of fragments with an appendage is invariable among samples, and mutations can be confidently judged by the appearance of bands which is absent in the lanes of normal DNA.

Additional merit of treatment with Klenow fragment is that the enzyme degrades unused primers by its 3' exonuclease activity. Primers at high concentrations may anneal to single-stranded DNA, and this can be another source of '*iso*-conformer' in some cases (14). Removal of primers may also serve to simplify the electrophoretogram of SSCP.

Protocol 3. Treatment of PCR product with Klenow fragment

Reagents
- Klenow solution: 10 mM Tris–HCl pH 8.3, 20 mM MgCl$_2$, 50 μM dNTP (see *Protocol 2*), 0.2 U/μl of Klenow fragment of DNA polymerase I (Boehringer Mannheim or New England Biolabs)
- Formamide dye: see *Protocol 2*

Method
1. Add 5 μl of Klenow solution to 5 μl of labelled PCR product from *Protocol 2*, and vortex.
2. Incubate at 37°C for 15 min.
3. Stop the reaction by adding 90 μl of formamide dye and vortex.

2.5 SSCP gel electrophoresis

Many physical factors influence the conformation of single-stranded DNA. Among them, temperature of the gel needs particular attention. Dramatic changes of relative mobilities of sequence variants in SSCP analysis have been demonstrated using perpendicular temperature-gradient gel electrophoresis (18). Use of a thin gel, such as the one used in sequencing, is important to minimize temperature rise by Ohmic heating. Water-jacketed electrophoretic apparatus (e.g. ATTO) has the advantages of precise control of gel temperature, independent of room temperature. Down to 10°C below room temperature can be attained by using a cooling plate attached to one side. More economically, strong blowing by two cross-flow fans (laminar flow fans), each about a power of 30 W, from both sides of a gel plate is effective in keeping gel at a constant (room) temperature. With less efficient air cooling (such as by usual turbulent fans), electrophoresis should be run at lower wattage, perhaps at 10–20 watts. A sensitive alarm of warming is 'smiling' of migration of the leading dye, bromophenol blue.

Electrophoresis in SSCP is often carried out using two or more conditions to increase sensitivity (4). Using 1 × TME buffer pH 6.8 (TME 6.8 gel, see *Protocol 4*), and running at 25 °C is the first choice because it detects most mutations. Additional mutations may be found in other conditions such as using the same buffer running at 20°C, or using 0.5 × TBE buffer and 5% glycerol (0.5 TBE–glycerol gel, see *Protocol 4*) running at 25 °C. The effect of glycerol is primarily to reduce the pH by complex formation between glycerol and borate ion. The pH of 0.5 TBE–glycerol gel is 7.7 while the pH of TBE is 8.3. The glycerol effect is not observed in buffers which do not contain borate. Lowering of the pH seems to have effects on mobility similar to, but not exactly the same as, lowering the temperature (19). Perhaps, by lowering pH, the charges of phosphates in the nucleic acid backbone are suppressed, and more nucleotide residues become involved in maintaining the tertiary structure.

Use of gel with low cross-linker (low ratio of *N,N'*-methylene *bis*-acrylamide to acrylamide) is important for efficient detection of mutations. Gel fibre with less cross-link is more flexible, and this flexibility may be important for conformation-sensitive retardation of migrating macromolecules. However, the exact reason is unknown.

Protocol 4.   Gel electrophoresis

Equipment and reagents

- Electrophoretic apparatus: see above
- 5 × TBE: 0.45 M Tris base, 0.45 M boric acid, 10 mM Na₂EDTA pH 7.8
- 10 × TME pH 6.8: 0.3 M Tris base, 0.35 M Mes, 10 mM Na₂EDTA pH 7.8 (Mes is 2(*N*-morpholino)-ethanesulfonic acid) (Sigma Chemical Company)

- Acrylamide solution: 49.5% acrylamide, 0.5% *N,N'*-methylene *bis*-acrylamide (Bio-Rad Laboratories)
- 50% glycerol
- 1.6% ammonium persulfate
- TEMED (*N,N,N',N'*-tetramethylenediamine)

Method

1. Assemble glass plates for the gel.

2. Mix following solutions in a 50 ml tube. Volumes are for a gel of 0.03 × 30 × 40 cm. Change them appropriately for gel of different dimensions.

   (a) TME 6.8 gel:

10 × TME pH 6.8	4.5 ml
Acrylamide solution	4.5 ml
1.6% ammonium persulfate	1.5 ml
H₂O to make	45.0 ml

   (b) 0.5 TBE–glycerol gel:

50% glycerol	4.5 ml
5 × TBE	4.5 ml
Acrylamide solution	4.5 ml
1.6% ammonium persulfate	1.5 ml
H₂O to make	45.0 ml

14

3. Add 45 μl of TEMED, mix, and immediately pour into gel plate assembly. Keep the plate at a horizontal position, insert flat side of a shark tooth comb to the depth of approximately 5 mm, and leave for at least 2 h.

4. Remove the comb, set gel plate to electrophoresis apparatus (attach aluminium plate if you are using air-cooling system), fill the reservoir with 1 × TME or 0.5 × TBE, depending on gel, and thoroughly rinse the top surface of gel with the buffer using a Pasteur pipette. Set the comb tooth-side down.

5. Heat the PCR product in formamide dye at 80°C for 5 min, and load onto the gel (1 μl per 5 mm lane).[a]

6. Start electrophoresis at 40 W. Also start cooling by circulating temperature-controlled water to the water-jacket, or by blowing with fans, depending on the instrument.[b]

7. Transfer the gel on a sheet of filter paper, cover with Saran Wrap, and dry using a gel dryer. Contact the dried gel to X-ray film and mark position of contact by stapling at three corners.

8. Expose for few hours to overnight depending on radioactivity in the PCR product. Remove staples and develop film.

[a] Quenching heated samples before loading onto the gel is not recommended, because it may encourage association of remaining primers and single-stranded DNA and lead to complication of final SSCP electrophoretogram (14). The shift of mobility of a mutated sequence may be subtle, and so for the best detection of mutations, load DNA of reference samples every three lanes so that the lanes of DNA of any test sample is adjacent to that of a reference.
[b] Time to stop the electrophoresis depends on length and sequence of the fragment. Suggested time for the first trial is when the bromophenol blue reaches 5 cm from the bottom for 150 bp fragments, and when xylene cyanol reaches 5 cm from the bottom for 400 bp fragments.

2.6 Interpretation of autoradiogram

Typically, autoradiogram of a single molecular species (e.g. cloned DNA, homozygous locus) should give a maximum of two bands, each for the separated strands. Only a single band should be observed if the two strands are not resolved. Without cleaning-up of the 3′ end, however, three or more bands may appear, depending on the sequence of the target and electrophoretic conditions. Most of these extra bands are explained by the heterogeneity at the 3′ end created during PCR. Appearance of a band which is absent in the reference lane indicates presence of a mutation.

2.7 Re-amplification and direct sequencing

Once the shifted bands are observed in the autoradiogram, they are excised as small pieces of dried gel, and used for extraction and re-amplification of DNA of single alleles, for subsequent cycle sequencing to determine the mutated nucleotides (8).

Protocol 5. Identification of mutations by sequencing

Equipment and reagents

- Apparatus and reagents for agarose gel electrophoresis: see *Protocol 1*
- Micro concentrator: Microcon 100™ (Amicon)
- Cycle sequencing kit: various kits are commercially available (e.g. AmpliCycle, PE Applied Biosystems; Thermo*Sequenase*, Amersham)

Method

1. Place filter paper carrying dried gel (gel-side up) on top of the developed X-ray film and align exactly at the place of contact by matching the holes of the staples. Fix the position again by stapling.

2. Place the film/filter paper/gel on a light box, and wet appropriate range of filter paper with ethanol to make the filter paper semi-transparent.

3. Cut a small area, e.g. 1 mm × 2 mm, of gel corresponding to the bands of interest (mutant and normal), using a razor blade. Remove Saran Wrap with fine forceps, then peel off dried gel from filter paper and drop into 20 μl H_2O in a 0.5 ml tube.

4. Heat the tube at 80°C for 3 min, allow to cool to room temperature, and briefly centrifuge to collect all the water at the bottom.

5. Take 1 μl of the water extract and subject to 25 cycles of PCR with 1 μM of each primer (the same primer set used in the first PCR) and 100–200 μM of each of the four deoxynucleotide triphosphates in 20 μl.

6. Add 200 μl of H_2O, apply to Microcon 100 micro concentrator, and centrifuge at 500 *g* for 10 min. Wash twice by adding 200 μl of H_2O to the retentate and centrifuging in the same manner.

7. Recover the retentate by fitting the inner cup at inverted position to a new bottom tube and centrifuging at 1000 *g* for 3 min. Adjust the volume to 20 μl by adding H_2O.

8. Estimate DNA concentration by agarose gel electrophoresis.[a]

9. Use 0.02–0.05 pmol of the cleaned-up, re-amplified DNA for cycle sequencing reaction. Kits are commercially available.

[a] See footnote in *Protocol 1*.

2.8 Gel matrices other than polyacrylamide

Gel matrices other than described here may enhance separation of mutated strands. Those are polyacrylamide gel at higher concentrations (20) or a new gel matrix, MDE™ gel (FMC) (21). The time required for electrophoresis is generally longer with these gels, and whether to use them may be a trade-off between efficiency and sensitivity. Agarose gel is a candidate gel matrix for separation of SSCP conformers of very long DNA fragments. Minisatellite isoalleles (same length but a few base substitutions within the minisatellite repeat) of as long as 6.3 kb could be successfully resolved using agarose gel (22).

2.9 Restriction endonuclease fingerprinting and dideoxy fingerprinting

Reduced sensitivity in long DNA fragments is a disadvantage of SSCP when searching for mutations in a long stretch of DNA. One solution to circumvent this problem and to efficiently search for mutations in long DNA segment is to amplify the target in a long PCR, digest the product with restriction enzyme, and analyse the mixture of fragments by SSCP (restriction endonuclease fingerprinting or REF) (see ref. 23). Using a set of a few restriction enzymes (usually of four base recognition), virtually all mutations can be detected. The power of this approach is demonstrated in the search of mutations in a 9 kb coding region of ATM mRNA from ataxia telangiectasia patients (24). This method may be more sensitive to noise, e.g. inadvertent amplification, partial digestion and annealing of strands, than simple SSCP. With all the necessity of the extra manipulation and care, however, REF may be an efficient method for the first screening of rare mutations in long, contiguous segments of DNA.

Among the efforts to improve sensitivity of SSCP is dideoxy fingerprinting, in which products of one of the four dideoxy sequencing reactions are analysed by SSCP gel electrophoresis (25). The major problem in the SSCP analysis is that the sensitivity is dependent on the sequence context in an unpredictable way. The dideoxy terminated DNA fragments that extend beyond the mutation points provide chances of separation of mutations in many sequence backgrounds, although they are not independent context in the strict sense. It is not surprising that the sensitivity of this method is virtually 100% for up to 600 bp of segments, if the dideoxy reaction products from both primers of the PCR segments are analysed (26). The major criticism against this method is that since it involves at least one dideoxy sequencing reaction, which is an extraneous manipulation in addition to PCR, the advantage over direct sequencing, that detects the exact mutated sequences, is not obvious.

3. Fluorescent SSCP in an automated DNA sequencer

Alternatives to radioisotopic labelling are fluorescent labelling, or staining by, for example, silver (27) after separation. The staining method is safe and convenient for testing for the presence or absence of mutations known to be detectable, but it is not suitable for searching for unknown mutations because of the low resolution and consequently low sensitivity, due to the use of a relatively short and thick gel.

Techniques to separate and detect fluorescently labelled DNA fragments at a high resolution are well established owing to the advancement of sequencing technology, and use of an automated DNA sequencer is a logical choice for non-isotopic high sensitivity SSCP (28). Fluorescently labelled DNA fragments can be easily obtained by using labelled primers in PCR, but chemical synthesis of fluorescence labelled oligonucleotides is costly. Recently, a simple and less costly procedure for post-PCR multicolour fluorescent labelling of amplified DNA has been developed (*Protocol 6*) (see also ref. 29).

In the procedures described below, an automatically operated capillary electrophoresis system, ABI PRISM 310 (PE Applied Biosystems) is used. In this system, DNA fragments labelled with one or more fluorophores are separated by electrophoresis in an uncoated fused silica capillary which is filled with a low viscosity polymer that has sieving properties (30). Separated fragments are excited by a laser beam near the cathode end of the capillary, and emitted fluorescence is spectrally separated by a grating mirror. Fluorescence is collected at short time intervals, 4.5 per sec, by an array of CCD, at wavelength ranges which are specified for the individual fluorophores. Fluorescence intensity of each fluorophore at the specified wavelength is recorded after correcting spectral overlap of other fluorophores.

Highly sensitive SSCP analysis by capillary electrophoresis has been reported (31, 32). The high sensitivity owes much to the high resolving power of the capillary electrophoresis. The present system has the capability of multi-colour detection, which allows each strand of a DNA fragment to be labelled with a different fluorophore. Moreover, the inclusion of an internal standard labelled with a third fluorophore aids in the compensation of run-to-run variations. The system is fully automated and skill-demanding manual gel casting and sample loading are no longer necessary. Most importantly, mobilities are normalized and digitized by the computer, and the presence or absence of mutations are judged by statistical evaluation. Many of the advantages except for gel casting and sample loading are shared with SSCP using gel-based, automated, multicolour DNA sequencers.

A disadvantage of this system is the considerable initial investment for the machine. Although the actual running time required for the analysis of one sample is approximately 20 minutes, the number of analysable samples is limited to about 60 per day per machine, due to the design of the system in which samples are serially analysed in a single channel capillary, and the sieving matrix in the capillary is replaced between each run. On the other hand, up to 96 samples can be analysed in an unattended operation.

3.1 Primer design in post-PCR fluorescent labelling

PCR products can be internally fluorescently labelled by including fluorescent nucleotides during the amplification reaction. However, such products are not suitable for SSCP, because the number and position of incorporated fluorescent nucleotides are different from molecule to molecule, and so, they are chemically heterogeneous. PCR products using primers fluorescently labelled at their 5′ ends are suitable for SSCP. However, chemical synthesis of fluorescent primers is expensive.

Alternatively, PCR products can be fluorescently labelled after the amplification reaction, by the 3′ exchange reaction of Klenow fragment of DNA polymerase I (29). Fluorophores of commercially available fluorescent nucleotides suitable for the instrument are R110, R6G, or TAMRA. These rhodamine derivatives are attached to position three of the pyrimidine rings of either dUTP or dCTP via amino-propynyl linker arms (33).

In order for those fluorescent nucleotides to be incorporated in the exchange reaction, PCR products must have C or T at their 3′ ends. This is achieved by using PCR primers having G or A at their 5′ ends. Also, purine nucleotides should not be at the 5′ penultimate positions of primers to avoid internal incorporation of fluorescent nucleotides. We usually design primers to have 5′-GTT or 5′-ATT, the second T as a precaution. These 5′ three nucleotides of primers need not be complementary to the template, since they do not critically affect the specificity of PCR, if the remaining parts of the primers that base pairs with the template are sufficiently long.

3.2 Fluorescent labelling by 3′ exchange reaction

Fluorescence labelled nucleotides are poor substrates for the 3′ exonuclease activity of Klenow fragment. And so, if the 3′ exchange reaction by the enzyme is carried out in the presence of fluorescence labelled and ordinary nucleotides, the fluorescent nucleotides remain unexchanged once they are incorporated into the 3′ ends. As a result, 3′ end-labelled DNA fragments accumulate during the exchange reaction (*Figure 2*). This labelling reaction can be carried out in the same tube used for PCR without any fractionation steps, simply by adding the appropriate reagents (29).

During the labelling reaction, the extra nucleotide which is often added at the 3′ ends during the PCR are automatically removed, and the labelled fragments have a homogeneous 3′ structure. Klenow fragment also degrades unused primers by its 3′ exonuclease activity. Removal of free primers assures that no primer:DNA interaction occurs during electrophoretic separation. After labelling, fluorescent nucleotides which interfere with electrophoretic analysis are degraded by adding alkaline phosphatase.

In *Protocol 6*, the fragment to be labelled is a PCR product using one primer having an A and the other having a G at their 5′ ends. One strand of the PCR product is labelled with R110-dUTP, and the other strand with R6G-dCTP at their 3′ ends by this method. The protocol can be modified so that both strands are labelled in the same colour. We differentially label PCR fragment of sample (suspected to be mutant) and reference (known to be

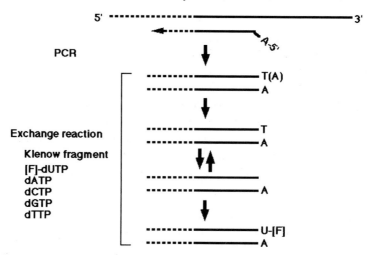

Figure 2. Illustrates how a PCR fragment is labelled with fluorescence at its 3' ends in the exchange reaction by the Klenow fragment of DNA polymerase I. A PCR product amplified with a primer having A at its 5' end has T at its 3' end, frequently with appended A. The product is treated with Klenow fragment in the presence of all four ordinary deoxynucleotides and [F]-dUTP (dUTP modified with fluorophore). The enzyme first removes appended A, then exchanges terminal T with T or [F]-U. Fragments with [F]-U at the 3' end accumulate because the terminal residue is not exchanged.

normal) DNAs. Reference DNA is also labelled with TAMRA in both strands, and used as an internal control in the electrophoresis as described in Section 3.3.

Protocol 6. Post-PCR 3' fluorescent labelling

Reagents

- 10 × Klenow buffer: 50 mM Tris–HCl pH 8.7, 0.1 M MgCl₂
- 0.2 M Na₂EDTA pH 7.8
- Formamide
- R110-dUTP, R6G-dCTP, TAMRA-dUTP, TAMRA-dCTP: dissolved in 30 mM Tris–HCl pH 9.5 and stored at –20°C (PE Applied Biosystems)

- Klenow fragment of DNA polymerase I: 5 U/μl (Boehringer Mannheim or New England BioLabs)
- Calf intestine alkaline phosphatase: 2 U/μl—a high concentrtrion (20 U/μl) of the enzyme (Boehringer Mannheim) is diluted with 50 mM Tris–HCl pH 8.5, 0.1 mM EDTA

Method

1. Amplify the target sequence by PCR using one primer having an A and the other having a G at their 5' ends (primers at 0.4–1 μM each, and nucleotides at 100–200 μM each).[a]

2. Combine the following solutions to make 50 μl of labelling mix. This is enough for ten labelling reactions.

- H_2O 34.0 μl
- 10 × Klenow buffer   10.0 μl
- 100 μM R110-dUTP   2.0 μl
- 100 μM R6G-dCTP   2.0 μl
- Klenow fragment   2.0 μl

3. Add 5 μl of labelling mix to each PCR tube containing 5 μl of amplified product, vortex, and incubate for 15 min at 37°C.[b]

4. Stop the action of the Klenow fragment by adding 1 μl of 0.2 M EDTA, then add 1 μl of calf intestine alkaline phosphatase, vortex, and incubate for 30 min at 37°C to degrade the nucleotides.

5. Add 90 μl of formamide and vortex.

[a] Successive addition of reagents is conveniently done if no mineral oil is overlaid. For this reason, a thermal cycler equipped with a heating lid is recommended.
[b] In typical PCR reactions using 100–200 μM of each of the four nucleotides, most of the nucleotides remain unincorporated at the end of the reaction when the amplification has reached a plateau. Consequently, in the 3′ exchange reaction described here, ordinary nucleotides are present at 25–50 molar excess of fluorescently labelled nucleotides.

3.3 SSCP in capillary electrophoresis (CE-SSCP)

In mutation finding by SSCP, electrophoretic mobilities of strands of PCR product from sample DNA is compared with those from a reference (usually normal or wild-type) DNA. Using the ABI PRISM 310 capillary electrophoretic system, samples are serially analysed, and mobilities in each run are calibrated using peaks of internal control DNA which are labelled in a colour different to those of sample DNA. Due to the algorithm of calibration, mobilities of the sample peaks are more reproducible if they are located closer to the peaks of the internal control. In the protocol that follows, TAMRA labelled PCR product from normal DNA is used as an internal control, together with TAMRA labelled size marker (PE Applied Biosystems). Mobility changes caused by different fluorophores are usually subtle or none, and peaks of strands of normal DNA labelled with R110 or R6G are closely located to the peaks of TAMRA labelled counterparts. So calibrated mobilities of normal DNA peaks are highly reproducible, and mutant DNAs which has even small mobility shifts can still be confidently detected.

The ABI PRISM 310 system allows for separations above ambient temperatures by the use of a heating plate that controls the temperature of most parts of the capillary. However, the range of temperature and the accuracy of control of this built-in system is inappropriate for SSCP analysis. We have evaluated subambient temperature separation with a prototype cooling device provided by PE Applied Biosystems. Though the temperature can be set independent of room temperature with this unit, it is safe to operate between 10°C above and 5°C below room temperature to obtain stable results and to avoid condensation.

Protocol 7. Separation in capillary electrophoresis

Equipment and reagents

- Automated capillary sequencer: ABI PRISM 310 (PE Applied Biosystems) equipped with a prototype cooling unit connected to a water-bath circulator kept at a constant temperature
- Capillary: uncoated capillary of 50 μm i.d., 61 cm (PE Applied Biosystems) is cut to 41 cm (this capillary has a lifetime of more than 100 electrophoretic runs without observable loss of resolution)
- Reservoir buffer: 1 × TBE containing 10% glycerol (see *Protocol 4* for the buffer)

- Size marker: GeneScan 2500 TAMRA or GeneScan 500 TAMRA (PE Applied Biosystems)
- Separation matrix: 3% GeneScan Polymer (PE Applied Biosystems) in 1 × TBE containing 10% glycerol (prepared according to manufacturer's instructions)
- Deionized formamide: see *Protocol 2* for deionization of formamide

Method

1. Combine the following solutions in 310 Genetic Analyzer sample tubes:[a]

   - Deionized formamide   13.0 μl
   - TAMRA labelled size marker   1.0 μl
   - TAMRA labelled normal PCR product   0.5 μl
   - R110 or R6G labelled test DNA   0.5 μl

2. Heat at 90°C for 3 min and place in autosampler tray.

3. Start electrophoresis at 20°C or 30°C.[b] Electrophoretic sample injection is at 15 kV for 5 sec, and separation is at 15 kV for 15 min.[c]

4. Begin data collection 5 min after the start of electrophoresis for fragments approx. 600 bp.

5. At the end of the separation, replace the matrix by pressurized injection from reservoir (glass syringe) for 2 min.

6. Load the next sample and start separation as in step 3.

[a] Volumes of TAMRA labelled normal DNA product and R110 or R6G labelled test DNA can be adjusted depending on signal intensity.

[b] Start water circulation at least 2 h before electrophoresis. The temperature of the blocks that sandwich the capillary reaches steady state rather slowly. The heater of the built-in heating plate should be turned off by setting the temperature below that of the water circulator. At least five of the test DNAs are devoted to normal DNA to statistically evaluate the variability of its mobility (see Section 3.4). Detailed operations (e.g. adjustment of tray movement, injection/replacement of matrix, installation of electrode and capillary) are found in the manual for the instrument. Washing the capillary with alkali, acid, or water has deleterious effects on resolution, and should not be done.

[c] The machine is usually set to 200 msec of data collection and 20 msec of interval for sequencing and minisatellite typing purposes. This should be changed to 50 msec for data collection and 20 msec for interval.

Like gel-based SSCP analysis, sensitivity in terms of per cent detection of mutations in CE-SSCP increases by running the electrophoresis in combinations of two or more conditions. By a single electrophoresis at either 20°C or 30°C, 22 out of 23 mutant fragments of 592 bp (all base substitution mutants) has been detected, whereas all were detected by a combination of the two temperatures. Thus, CE-SSCP seems to be more sensitive than conventional gel-based SSCP using TBE–glycerol buffer. But it is still too early to generalize about the sensitivity of CE-SSCP, because the data available are limited to a few sequence contexts.

3.4 Data processing

Results of runs are obtained as raw data, which can be displayed graphically with fluorescence intensity in the ordinate and data points in the abscissa. Raw data are corrected for variations of mobility among runs. Corrected data are then used to evaluate the variability of mobility of reference DNA, in this case normal DNA. Sample DNA showing mobility outside variation of mobility of normal DNA is judged to be mutant.

Mobility correction is made first by choosing one of the runs as a template, and then by aligning other runs to this template referring to peaks of internal control DNA which, in this case, is labelled with TAMRA. Exact normalization of mobilities (expressed in data point) of sample DNA (labelled with R110 or R6G) is made by 'local Southern' method (34). This method requires defining at least two reference (TAMRA) peaks on either side of the sample (R110 or R6G) peaks. Two of the references are peaks of TAMRA labelled reference DNA, and others are peaks of the size marker labelled with TAMRA.

Mobilities of reference and sample DNA (labelled in R110 or R6G) are tabulated in data point, and the values are exported to a database, such as Microsoft Excel. Standard deviations of mobilities of peaks of normal DNA are then calculated, and sample DNA which shows mobility outside three times the standard deviation in at least one strand is judged to be mutant.

Acknowledgements

We thank Drs H. Michael Wenz, Margaret Galvin, Marianne Hane, Michael Phillips, and other people in PE Applied Biosystems for collaboration, especially by making available an ABI PRISM 310 and a prototype of the cooling unit for the capillary electrophoresis system.

References

1. Orita, M., Suzuki, Y., Sekiya, T., and Hayashi, K. (1989). *Genomics*, **5**, 874.
2. Glavac, D. and Dean, M. (1993). *Hum. Mutat.*, **2**, 404.
3. Kukita, Y., Tahira, T., Sommer, S. S., and Hayashi, K. (1997). *Hum. Mutat.*, in press.

4. Hayashi, K. and Yandell, D. W. (1993). *Hum. Mutat.*, **2**, 338.
5. Fischer, S. C. and Lerman, L. S. (1983). *Proc. Natl. Acad. Sci. USA*, **80**, 1579.
6. White, M. B., Calvalho, M., Derse, D., O'Brien, S. J., and Dean, M. (1992). *Genomics*, **12**, 301.
7. Winter, E., Yamamoto, F., Almoguera, C., and Perucho, M. (1985). *Proc. Natl. Acad. Sci. USA*, **82**, 7575.
8. Cotton, R. G. H., Rodrigues, N. R., and Campbell, R. D. (1988). *Proc. Natl. Acad. Sci. USA*, **85**, 4397.
9. Suzuki, Y., Sekiya, T., and Hayashi, K. (1991). *Anal. Biochem.*, **192**, 82.
10. Suzuki, Y., Orita, M., Shiraishi, M., Hayashi, K., and Sekiya, T. (1990). *Oncogene*, **5**, 1037.
11. Gaidano, G., Ballerini, P., Gong, J. Z., Inghirami, G., Neri, A., Newcomb, E. W., *et al.* (1991). *Proc. Natl. Acad. Sci. USA*, **88**, 5413.
12. Cawthon, R. M., Weiss, R., Xu, G., Viskochil, D., Culver, M., Stevens, J., *et al.* (1990). *Cell*, **62**, 193.
13. Hayashi, K. (1995). In *Laboratory protocols for mutation detection* (ed. U. Landegren), p. 14. Oxford University Press, Oxford.
14. Cai, Q-Q. and Touitou, I. (1993). *Nucleic Acids Res.*, **21**, 3909.
15. Sambrook, I., Fritch, E. F., and Maniatis, T. (ed.) (1982). *Molecular cloning: a laboratory manual*, p. 400. Cold Spring Harbor Laboratory Press, NY.
16. Kellog, D. E., Rybalkin, I., Chen, S., Mukhamedova, N., Vlasik, T., Siebert, P. D., *et al.* (1994). *BioTechniques*, **16**, 1134.
17. Sugano, K., Nakashima, Y., Yamaguchi, K., *et al.* (1996). *Genes, Chromosomes and Cancer*, **15**, 157.
18. Sugano, K., Fukayama, N., Ohkura, H., Shimosato, Y., Yamada, Y., Inoue, T., *et al.* (1995). *Electrophoresis*, **16**, 8.
19. Hayashi, K. (1992). *Genetic analysis: techniques and applications*, **9**, 73.
20. Savov, A., Angelicheva, D., Jordanova, A., Eigel, A., and Kalaydjieva, L. (1992). *Nucleic Acids Res.*, **20**, 6741.
21. Keen, J., Lester, D., Inglehearn, C., Curtis, A., and Bhattacharya, S. (1991). *Trends Genet.*, **7**, 5.
22. Monckton, D. G. and Jeffreys, A. (1995). *Nucleic Acids Res.*, **22**, 2155.
23. Liu, Q. and Sommer, S. (1995). *BioTechniques*, **18**, 470.
24. Gilad, S., Khosravi, R., Shkedy, D., *et al.* (1996). *Hum. Mol. Genet.*, **5**, 433.
25. Sarkar, G., Yoon, H., and Sommer, S. S. (1992). *PCR Methods Appl.*, **4**, 97.
26. Liu, Q., Feng, J., and Sommer, S. S. (1996). *Hum. Mol. Genet.*, **5**, 107.
27. Ainsworth, P. J., Surh, L. C., and Coulter-Mackie, M. B. (1991). *Nucleic Acids Res.*, **19**, 405.
28. Makino, R., Yazyu, H., Kishimoto, Y., Sekiya, T., and Hayashi, K. (1992). *PCR Methods Appl.*, **2**, 10.
29. Inazuka, M., Tahira, T., and Hayashi, K. (1996). *Genome Res.*, **6**, 551.
30. Wenz, H. M. (1994). *Nucleic Acids Res.*, **22**, 4002.
31. Hebenbrock, K., Williams, P. M., and Karger, B. L. (1995). *Electrophoresis*, **16**, 1429.
32. Kuypers, A. W., Williams, P. M., van der Schans, M. J., Linssen, P. C., Wessels, H. M., de Bruijin, C. H., *et al.* (1993). *J. Chromatogr.*, **621**, 149.
33. Lee, L. G., Connel, C. R., Woo, S. L., *et al.* (1992). *Nucleic Acids Res.*, **20**, 2471.
34. User's manual. (1995). *ABI PRISM GeneScan Analysis Software*. PE Applied Biosystems, Foster City, CA.

2

Single-stranded conformation polymorphism and heteroduplex analysis

BERNARD GERRARD and MICHAEL DEAN

1. Introduction

Single-stranded conformation polymorphism (SSCP) alone or in combination with heteroduplex analysis (HA) is one of the most widely used and practical approaches for mutation detection (1). The principal advantage of this method is that it is rapid to perform and can be carried out using equipment available in most molecular biology laboratories. When properly optimized, the method is highly sensitive, and many different mutations within a DNA fragment can often be distinguished on the same gel (2, 3). SSCP and HA can be performed on the same gel since, after the denaturation of the sample prior to loading, there is often the re-formation of a significant amount of double-stranded DNA which will appear in a lower position from the single-stranded products on the SSCP gel. Since heteroduplexes can often be resolved from homoduplexes, the appearance of heteroduplexes on the SSCP gel can give additional information on the presence of variants.

In SSCP/HA analysis the sample is amplified by the polymerase chain reaction (PCR) in the presence of a radiolabelled nucleotide, usually ^{32}P. This product is denatured to generate single-stranded molecules and loaded on a non-denaturing gel. The single-stranded molecules are resolved on the gel, and DNA sequence alterations between the primers appear as fragments of altered mobility due to differential folding of the single strand. The conditions of the gel can be varied by altering the running temperature, the degree of cross-linking in the gel matrix, or by the inclusion of glycerol or sucrose. These variations change the type of conformations seen and can increase the sensitivity of detection (4). *Figure 1* displays analysis of several point mutations in the chemokine receptor 5, *CCR5*, gene. Variations are observed both in the upper (ss) regions and the lower (ds) portion of the gel. Lanes 2, 4, 8–11 contain the same alteration and give identical SSCP and heteroduplex patterns. Lanes 3 and 12 contain unique alterations that can be distinguished

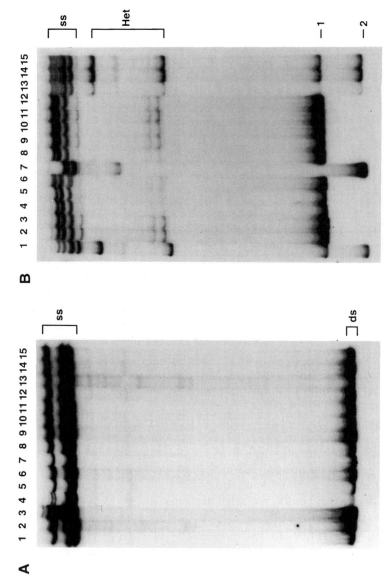

Figure 1. SSCP detection of multiple mutations. (A) Analysis of the 154–558 bp region of the chemokine receptor 5 (*CCR5*) gene (GenBank No. X91492). Samples were amplified with the CCR5F3 and CCR5R3 primer (26) and run on a 6% acrylamide gel at 4°C, 50 W. Lanes 1, 5–7, 13–15 are wild-type samples, and lanes 2–6, 8–12 have point mutations. ss, single-stranded DNA; ds, double-stranded DNA. (B) Analysis of a 32 bp deletion in *CCR5*. The 587–993 bp region was amplified with primers CCR5f4 (5′ CTTCTTCATCATCCTCCTGACA) and CCR5F4 (5′ TGTAGGGAGCCCAGAAGAGA) and run as described for panel A. Lanes 1, 13–15 contain heterozygotes for the CCR5Δ32 allele (26) and lane 7 a homozygote. Het, heteroduplex.

26

in either the ss or ds regions. Lanes 1, 5–7, 13–15 contain samples that are wild-type in this region.

Figure 1B displays analysis of samples that are heterozygous (lanes 1, 13–15) or homozygous (lane 7) for a 32 bp deletion in *CCR5*. The deletion can be visualized as a higher mobility fragment in the ds region. Note that the heteroduplexes (present only in the heterozygotes) are greatly retarded and nearly overlap the ss region. SSCP alterations are also observed. Insertions and deletions do not always result in dramatic SSCP changes, but can nearly always be detected as heteroduplexes.

2. Optimization of the PCR reaction

Optimum conditions for the PCR are essential to generating easily interpretable SSCP gels (see also Chapter 1). Important variables to consider are the concentration of Mg^{2+} in the reaction and the parameters of the PCR cycles in terms of temperatures and the number of cycles. We have recently found that a step-down PCR program gives clean product from a number of primers using identical Mg^{2+} concentrations (5). In this procedure, the annealing is initially carried out at a high temperature, often above the T_m of the primers, and gradually lowered. In this program the reaction is initiated at the temperature that is optimum for the fully base paired product. In subsequent cycles at lower temperatures, a sufficient mass of the expected product is generated, outcompeting any unwanted side-reactions. This procedure allows primers to initiate priming at their optimum temperature, eliminating the need for optimization of Mg^{2+} concentrations. An example of this type of program is given in *Protocol 1*.

Protocol 1.   PCR set-up and optimization

Equipment and reagents

- PCR machine
- Small (10–20 cm) acrylamide or agarose gels for visualizing PCR products
- 100–500 V power supply
- UV light box with camera for photographing gels

Method

1. Make a cocktail of all of the following reagents except the DNA. Add the DNA to the PCR tube and then add 24 μl of the cocktail and one drop of mineral oil. Spin briefly in a microcentrifuge and place in the PCR machine. For a 25 μl PCR reaction:

   - 1.0 μl genomic DNA (100 ng)
   - 2.5 μl 10 × PCR buffer
   - 2.5 μl 200 μM dATP, dCTP, dGTP, dTTP solution
   - 1.0 μl primer 1 (1 OD/ml)

Protocol 1. *Continued*

- 1.0 μl primer 2
- 16.8 μl water
- 0.2 μl *Taq* polymerase (5 U/μl)

2. A standard PCR program is 94°C for 3 min. Followed by 35 cycles of: 94°C 0.5 min, 55°C 0.5 min, 72°C 1.5 min, with a final extension of 72°C for 10 min.

3. The step-down program is 94°C for 3 min. Then two cycles of: 94°C 30 sec, 65°C 30 sec, 72°C 1 min. Then two cycles of: 94°C 30 sec, 64°C 30 sec, 72° C 1 min. Continue stepping down 1°C for the annealing step, two cycles each step, until 59°C is reached. Then 22 cycles of: 94°C 30 sec, 50°C 30 sec, 72°C 1 min, with a final extension at 72°C for 10 min.

4. Following the PCR, run 10 μl of the reaction on an 8% acrylamide gel or a 1.5% agarose gel. The conditions giving the highest yield and cleanest product will be used for SSCP analysis.

3. SSCP sample preparation

Once optimum conditions for the PCR are established, a scaled down re-action (10 μl) is performed with the presence of a radiolabelled nucleotide, typically ^{32}P. Other labelled isotopes such as ^{35}S or ^{33}P can be used. Alternatively, the primers can be end-labelled using γATP. (Note that both primers need to be labelled to visualize the two strands, since both strands must be visualized in SSCP analysis because conformers can be observed on either strand.) Amplified samples are then run on a long, thin acrylamide gel.

Protocol 2. Preparing PCR samples for SSCP/HA

Reagents
- Radiolabelled dCTP

Method

1. The SSCP reaction is identical to the test reaction except that 0.1 μl of [^{32}P]dCTP is added to the reaction, and the final volume is 10 μl. For a 10 μl labelled SSCP-PCR reaction use:

- 1.0 μl genomic DNA (100 ng)
- 1.0 μl 10 × PCR buffer
- 0.3 μl 200 μM dATP, dCTP, dGTP, dTTP solution
- 1.0 μl primer 1 (1 OD/ml)

- 1.0 μl primer 2
- 0.1 μl [^{32}P]dCTP (3000 Ci/mmol)
- 5.5 μl water
- 0.1 μl *Taq* polymerase (5 U/μl)

2. 2 μl of the PCR product is mixed in a well of a microtitre plate with 8 μl of loading dye (95% formamide, 0.1% bromophenol blue, 0.1% xylene cyanol, 10 mM NaOH). Alternatively 30 μl of loading dye can be added to the entire sample. Heat the plate or tubes to 95°C for 3 min.

3. Load 2–3 μl of the sample onto the gel. It is also useful to load a lane of undenatured DNA to display the double-stranded DNA fragments.

Protocol 3. Preparing an SSCP/HA gel

Equipment and reagents

- 35 cm sequencing gel box with 0.4 mm spacers and shark's tooth combs
- 2000–3000 volt power supply
- 4°C cold room

- 40% acrylamide:*bis* (37.5:1) (2.6% C) solution: dissolve 39.5 g acrylamide and 1.06 g *bis*-acrylamide in a total volume of 100 ml of distilled water

Method

1. Treat one glass plate with Repel Silane. This treatment needs to be done only occasionally.

2. Clean a set of plates for a 35 cm, 0.4 mm thick gel.

3. Prepare one of the gel formulations (listed below) and use to prepare the gel. For 75 ml of gel solution add 500 μl of 10% ammonium persulfate solution and 50 μl of TEMED. Allow the gel to polymerize for 30 min., and pre-run in the gel box for at least 5 min.

4. Load 2–3 μl of sample and allow to run until the bromophenol blue has reached the bottom.

5. Carefully separate the plates and remove the gel onto a piece of Whatman 3MM filter paper and cover with plastic film. Dry the gel and expose to X-ray film for 2–24 h at −70°C or at room temperature.

---

4. Optimization of SSCP/HA detection

The conditions of the SSCP gel can be varied to produce gels in which the samples form alternative conformations. The number of possible conformations is so large that there is no theoretical basis for choosing conditions. Clearly, the more conditions run, the greater the sensitivity. The researcher, however, is limited in the number of gels that can be run. A single gel can give

60–90% efficiency of detection and can be used as an initial screening method. Two to four different gels can be employed to reach close to 100% detection. The variables that have been employed are as follows:

(a) Temperature:
  i. room temperature overnight at 15 W constant power for a total of 7000–9000 volt hours
  ii. 50 W 3–5 h in a 4°C cold room

(b) Cross-linking:
  i. 19:1 (5% C)
  ii. 37.5:1 (2.6% C)
  iii. 75:1 (1.3% C)
  iv. 0.5 × MDE (Avitech Diagnostics)

(c) Additives:
  i. 5–10% glycerol
  ii. 10% sucrose

Each of these variables can alter the conformation of single-stranded molecules. SSCP was initially run on 5% acrylamide gels with 5% *bis*-acrylamide cross-linker (5% C) at either room temperature or 4°C with or without glycerol. More recently researchers have shown that gels with higher percentage of acrylamide and lower cross-linking can detect more mutations (6). Sucrose has also proven to be a useful additive (4). MDE is an alternative matrix that has been employed in place of acrylamide (4). *Table 1* gives gel recipes for several different conditions that produce optimum results (2). Unfortunately it is not possible to predict how a given change in conditions will affect the mobility of a specific fragment. Differences in the percentage of acrylamide and the extent of cross-linking affect the mobility of different mutations (2–4, 6).

Table 1. Gel formulations

Final	Stock	250 ml[a]
10% acrylamide (1.3% C)	40%	62.5 ml
1 × TBE	10 ×	25 ml
0.5 × MDE	2 ×	62.5 ml
10% glycerol	100%	25 ml
1 × TBE	10 ×	25 ml
10% acrylamide (1.3% C)	40%	62.5 ml
10% sucrose	–	25 g
1 × TBE	10 ×	25 ml

[a] All solutions are to made up to final volume (250 ml) with distilled water.

The size of the PCR product can also affect sensitivity. In general, the smaller the product, the higher the sensitivity, with the optimum being 200–300 bp. SSCPs have been observed on products as small as 50 bp and as large as 1 kb. Longer products can be cleaved with restriction enzymes to yield a series of bands that can be assayed simultaneously.

5. Multiplexing

The following multiplexing methods are useful when a large number of samples are to be analysed. In each case a test gel with known sample sizes and mutations should be run to provide a visual comparison for the expanded project.

(a) Use two primer pairs for each PCR reaction, provided the amplified products differ in size by at least 75 base pairs. Pre-test the combinations since some primer pairs are incompatible with each other.

(b) Amplify fragments that differ in size by at least 75 base pairs or more separately and load both amplifications of the same DNA in the same well. Again, run a test lane of each fragment size separately to obtain a visual comparison.

(c) To screen many DNA samples, amplify two DNAs together, thereby halving the number of gels that need be run. If a lane appears with an alteration, the two DNAs that this lane represents will have to be amplified separately to identify the mutant.

(d) Some PCR machines have a 96-well microtitre plate capability. This is a quick and economical way of amplifying large numbers of samples, particularly if the panel of DNAs are to be analysed with a number of primer pairs. Make a master plate of the DNAs; the amount of DNA per PCR reaction should be diluted to 5 µl with water multiplied by the number of plates that will eventually be run. Use a multichannel pipettor to transfer 5 µl of the diluted DNA from the master plate to the plates that will be used in the PCR machine. Allow these samples to dry in the plates. Add 10 µl of the appropriate PCR reaction mix, plus a drop of oil, to each well and proceed with the amplification program. 30 µl of loading dye can be added to each well after cycling, using a multichanneled pipettor.

6. Interpretation of results

The interpretation of the results of an SSCP gel requires some experience. Controls with known mutations can assist in the analysis. Some double-stranded DNA often re-forms after the denaturation step which will produce bands nearer the bottom of the gel. It is helpful to include an undenatured sample to help in identifying these bands. Double-stranded heteroduplex molecules formed when a mutant and wild-type strand anneal, can also indicate sequence alterations. These heteroduplexes migrate just above the double-stranded DNA. While heteroduplexes are most prominent in the case of insertions or deletions, they can be seen with some point mutations (7).

Since the exact migration of SSCP conformers can vary from gel to gel, it is useful to include any available control samples at least once on each gel. While it is expected that the two strands will migrate differently, this is not always the case. Similarly, in a heterozygote, not all four strands are always resolved. It is important to note that a mutant sample should give a clearly different pattern from a wild-type sample. In addition, a heterozygote should display approximately equal intensity in all four bands.

Occasionally PCR artefact bands occur that can confuse the interpretation. If a band appears only in samples that amplified better than the rest, it is likely to be an artefact. In samples that amplify well or are overloaded, alternative conformations can appear. These may be the result of two conformers partially annealing. Diluting the samples will chase the double-stranded DNA into single strands and remove some of these additional bands. Where possible, confirmation of the segregation of altered bands should be demonstrated within a family. The final proof of the alteration ultimately comes from sequencing.

7. Applications

Since the invention of the PCR-SSCP technique in 1989, many applications of the method have been demonstrated. SSCP has been used to identify alterations in tumour DNA samples as well as mutations in human disease genes and their animal counterparts (1, 8–11). SSCP can be applied to RNA or to reverse transcribed RNA (cDNA) either to increase sensitivity or to assay multiple exons simultaneously (12, 13).

SSCP has been applied to complex loci with many alleles, such as the HLA locus (14). Since different alleles usually give distinct patterns, five to ten alleles can usually be resolved on the same gel. SSCP can also be readily applied to the mapping of genes in interspecies backcrosses, where differences between the parental species can be easily detected in a 200–500 bp PCR product. SSCP alterations are often found in the 3' untranslated regions and introns of genes (15, 16). The 3' untranslated regions are rarely disrupted by introns, and such polymorphic sequence-tagged sites can allow a gene to be mapped genetically in families, as well as physically on yeast artificial chromosome clones and radiation hybrids (17).

8. Other methods

SSCP using radiolabelled samples is an excellent method for many applications. However, other methods may be warranted for some applications. A successful 'cold' SSCP method has been developed (18). Denaturing gradient gel electrophoresis (DGGE) involves resolving heteroduplex molecules by running them through an increasing concentration of denaturant (19). This method detects nearly 100% of all mutations and is also performed 'cold' without using radioactive compounds. In the G,C clamped version of this method (20) a special primer is required. Excellent protocols are provided in ref. 21. Other established methods to consider are chemical cleavage and RNase protection (reviewed in ref. 21). Recently, methods to identify unknown mutations relying on enzymatic detection (enzyme mismatch cleavage, mismatch repair enzyme cleavage) (22–24) or high performance liquid chromatography (25) have been developed that may prove useful in population-based studies.

References

1. Orita, M., Suzuki, Y., Sekiya, T., and Hayashi, K. (1989). *Genomics*, **5**, 874.
2. Glavac, D. and Dean, M. (1993). *Hum. Mutat.*, **2**, 404.
3. Glavac, D. and Dean, M. (1995). *Methods Neurosci.*, **26**, 194.
4. Ravnik-Glavac, M., Glavac, D., and Dean, M. (1994). *Hum. Mol. Genet.*, **3**, 801.
5. Hecker, K. H. and Roux, K. H. (1996). *BioTechniques*, **20**, 478.
6. Ravnik-Glavac, M., Glavac, D., Chernick, M., di Sant'Agnese, P., and Dean, M. (1994). *Hum. Mutat.*, **3**, 231.
7. White, M. B., Carvalho, M., Derse, D., O'Brien, S. J., and Dean, M. (1992). *Genomics*, **12**, 301.
8. Claustres, M., Laussel, M., Desgeorges, M., Giansily, M., Culard, J-F., Razakatsara, G., *et al.* (1993). *Hum. Mol. Genet.*, **2**, 1209.
9. Soto, D. and Sukumar, S. (1992). *PCR Methods Appl.*, **2**, 96.
10. Dean, M., White, M. B., Amos, J., Gerrard, B., Stewart, C., Khaw, K-T., *et al.* (1990). *Cell*, **61**, 863.
11. Claustres, M., Gerrard, B., Kjellberg, P., Desgeorges, J., and Dean, M. (1992). *Hum. Mutat.*, **1**, 310.
12. Danenberg, P. V., Horikoshi, T. M., Volkenandt, M., Danenberg, K., Lenz, H. J., Shea, C. C. L., *et al.* (1992). *Nucleic Acids Res.*, **20**, 573.
13. Kozlowski, P., Sobczak, K., Napierala, M., Wozniak, M., Czarny, J., and Krzyzosiak, W. J. (1996). *Nucleic Acids Res.*, **24**, 1177.
14. Carrington, M., Miller, T., White, M., Gerrard, B., Stewart, C., Dean, M., *et al.* (1992). *Hum. Immunol.*, **33**, 208.
15. Nielsen, D. A., Dean, M., and Goldman, D. (1992). *Am. J. Hum. Genet.*, **51**, 1366.
16. Glavac, D., Ravnik-Glavac, M., O'Brien, S. J., and Dean, M. (1994). *Hum. Genet.*, **93**, 694.
17. Poduslo, S., Dean, M., Kolch, U., and O'Brien, S. (1991). *Am. J. Hum. Genet.*, **49**, 106.
18. Hongyo, T., Buzard, G., Calvert, R., and Weghorst, C. (1993). *Nucleic Acids Res.*, **21**, 3637.
19. Myers, R. M., Maniatis, T., and Lerman, L. S. (1987). In *Methods in enzymology*, Vol. 155, p. 501 (ed. R. Wu), Academic Press, San Diego.
20. Sheffield, V. C., Cox, D. R., Lerman, L. S., and Myers, R. M. (1989). *Proc. Natl. Acad. Sci. USA*, **86**, 232.

3

Comprehensive mutation detection with denaturing gradient gel electrophoresis

L. S. LERMAN and CHERIF BELDJORD

1. Introduction

Since the basic laboratory procedures for denaturing gradient electrophoresis and the introduction of G,C clamps have not changed significantly from previous descriptions, and they depart only slightly from conventional gel electrophoresis, we provide guidance and references to published complete instructions. However, the computational preparations for mutation searching are critical to efficient application of the method. We provide a detailed introduction to the computational tools, their use in planning a search, and their role in interpreting results.

In addition, we provide a detailed laboratory protocol for psoralen cross-linking, an effective alternative to G,C clamping, and description of a number of related methods with somewhat different advantages. Citations and references are presented for useful techniques as samples of the literature, not as a comprehensive review.

1.1 The scope of DGGE, its distinctive capabilities, and the nature of results

Denaturing gradient gel electrophoresis (DGGE) is a generic designation of a family of related methods for mutation detection, all based on the reduction in electrophoretic mobility of a DNA molecule in a densely obstructed medium when part of the double helix unravels. They offer *a priori* assurance that any difference between two sequences, wild-type and putative mutant, can be detected. This comprehensive detection applies to nearly all sequences in the genome. The procedure is non-destructive; if a segment of DNA is found to be of interest, it can be recovered, amplified, run again in a gradient again (or otherwise), and sequenced. Only simple, low cost, low-tech equipment is needed. Strand reassortment between mutant and wild-type molecules generates a distinctive quadruplet pattern. Base modifications, such as methylation, are also detected when DNA is examined without amplification or cloning. Methods in the DGGE family are equally useful for somatic and germline variants.

The term, DGGE, is taken here to include all gel electrophoretic methods in which separation depends on the difference in equilibrium denaturing conditions required to effect partial melting of DNA molecules of different sequence or those containing helical irregularities. Partial melting results in a substantial reduction in the electrophoretic mobility, relative to that of the intact double helix, at a critical denaturing condition. In addition to the eponymous scheme, migration at constant temperature into ascending concentration of a denaturing solvent, DGGE will be understood here to include constant denaturant gel electrophoresis (CDGE), thermal gradient gel electrophoresis (TGGE), thermal denaturing gradient gel electrophoresis (TDGGE), constant denaturant capillary electrophoresis (CDCE), two-dimensional DNA electrophoresis, temporal temperature gradient electrophoresis (TTGE), temperature sweep gel electrophoresis (TSGE), etc.

All have the following characteristics in common:

(a) Sensitivity to sequence differences is based on the dependence of thermal stability of the double helix on its sequence.

(b) Prototype and variant molecules are physically separated.

(c) Separated molecules can be recovered unchanged.

(d) Separation is due to the sensitivity of electrophoretic mobility in a densely obstructed medium to partial disordering (melting) of the helix. The medium may be either a gel or a solution of a linear polymer.

(e) Separation occurs under conditions of near equilibrium between the helical and the partially melted forms. Those conditions consist of a combination of temperature and a substantial concentration of a denaturant. Melting equilibrium must be distinguished from the conventional 'denatured' state, observed subsequent to melting in an environment in which the helix is stable.

(f) Since strands do not fully separate, the equilibrium remains unimolecular, independent of concentration.

(g) Transition through the helix–disorder equilibrium is usually effected by some combination of temperature and the presence of a denaturing solvent. For some solvents, these are linearly interchangeable.

(h) Sensitivity to sequence variation is predictable in practical detail by means of thermodynamic and electrophoretic theory.

(i) The identity or difference between mutations giving similar initial patterns can be established unequivocally.

The various members of the DGGE family offer different benefits and conveniences. Use of a gradient of denaturant, either solvent (DGGE) or temperature (TGGE, TDGGE), makes closely defined preliminary adjustment of conditions to the requirements of each sequence unnecessary, and the running time of the gel is not critical. However, mobility discrimination under uniform denaturant (CDGE), results in larger band separations, but the sharpening of bands in a gradient is lost. In a capillary the possibility of high field strength shortens the time required many-fold, no gel handling is needed, and data are recorded during the run. Allowing the temperature to rise during migration, the temperature ramp (TTGE), permits mobility separations without need for critical temperature determination, but the separation intervals are smaller than in uniform denaturant. Two-dimensional gels allow scrutiny of a large number of fragments with a considerable saving in manipulative effort, and PCR amplification can be highly multiplexed, rather than requiring a separate reaction for each segment.

Because these methods provide nearly certain detection of sequence variants, they have been useful in localizing genotypic variants in DNA segments small enough for easy sequencing. For example, using DGGE, 308 putative mutations were distributed in the phenylalanine hydroxylase gene, causing phenylketonuria, into 12 of the 13 exons of the gene; five additional mutations were found in splice sequences (1). Of the total of 308 mutations in their collection, only three were not identified—a detection yield of 99%. These methods are also appropriate for routine genetic analysis in a clinical environment (2).

Fragments can be recovered unchanged after separation by DGGE, and they can be used again for further analysis. For example, a mutant band homoduplex can be cut from the gel, eluted according to classical conditions, and amplified again by PCR. Where a weak band from a small amount of a variant is expected but submerged in the tail of a much stronger adjacent band, DNA extracted from the adjacent band can be re-amplified and subjected again to DGGE, permitting detection of 10^{-6} presence of the variant (3). Detection of somatic mutations often proceeds from mixed cell types, in which the mutant may be a minority.

The products of reassociating the strands of a pair of molecules differing by a single base pair form a characteristic quadruplet band pattern in the denaturing gradient (4). The patterns for a set of thalassaemia mutants in a 457 bp fragment of the human β-globin gene are shown in *Figure 1*. Each lane carries the product of heating and reassociating a mixture of PCR amplified DNA of a mutant with a single base substitution and of normal β-globin. The four different products of reassociation are almost always resolved, with the parental homoduplexes at the bottom. Where the sequence difference is a simple transition, AT ⟷ GC, the order of the bands from denser denaturant to lighter (bottom upward in the gel) is GC homoduplex, AT homoduplex, GT heteroduplex, and AC heteroduplex. The separation of the homoduplex and heteroduplex bands is usually larger than that within either pair.

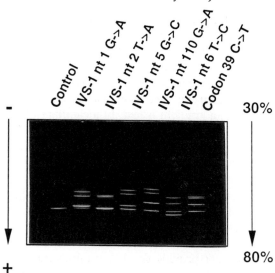

Figure 1. Denaturing gradient gel patterns from fragments of the human β-globin gene, bp 97–553, amplified from DNA of patients heterozygous for β-thalassaemia using a psoralen labelled primer at the 5′ end. The strands were separated and reassociated by heating and cooling, then irradiated for cross-linking. Base substitutions and positions as follows: lane 1, normal β-globin only; lane 2, bp 359 G→A, IVS1; lane 3, bp 360 T→A, IVS1; lane 4, bp 363, G→C, IVS1; lane 5, bp 469, G→A, IVS1 (110); lane 6, bp 364, T→C, IVS1. Denaturant (top 30%, bottom 80%) urea–formamide stock (7 M urea, 40% (v/v) formamide, TAE). Electrophoresis: 6 h at 60°C, 10 V/cm in 8% polyacrylamide. Stained with ethidium. The position numbering follows ref. 12.

2. Background

Our early studies showed that the mobility of a DNA molecule migrating in a gel containing a uniform concentration of a denaturing solvent is an intricate function of the denaturant concentration at temperatures between 50°C and 90°C. At critical denaturant concentrations, different and characteristic for each sequence, the mobility drops to a much lower value (5). We showed later that the temperature–denaturant concentrations giving the mobility transition among fragments of lambda DNA changed with single base pair substitutions in the fragments. At a denaturant concentration intermediate between the transitions of wild-type and mutant there is a large difference in their mobilities (6) (Figure 3 of that paper). These observations evolved into a more easily applicable procedure, migration into a gradient, where the result is given by the gradient level of the mobility change, not by the mobility itself.

We combined the gradient scheme with a prior orthogonal separation to give two-dimensional patterns, where the spot positions reflect both length of the fragments and an aspect of their thermal stability (5).

The resolving power of the process, which derives from the sequence dependence of the thermal stability of DNA, permits detection of single base changes. We showed that all base substitutions in one section of the lambda CII gene resulted in shifts in gradient retardation level, up or down, depending on the specific change, while substitutions in an adjacent section, 2 bp removed, did not effect shifts. Shifts were found even where the substitutions left the local overall base composition unaltered (6). The observations were in precise agreement with predictions based on statistical mechanical melting theory. The introducton of G,C-dense, thermally stable end sequences, 'clamps' (7, 8) permitted extension of the search capability to virtually all parts of the genome. Psoralen cross-linking at the end of the molecule, discussed below, fulfils the same function.

3. Basic principle, the physical properties of DNA

The full development of DGGE methods has come from understanding the relation between the helix–disorder equilibrium and base sequence. At high 'equivalent temperature' (the combined effect of denaturant and thermometer temperature) the two strands of the double helix completely dissociate. Within DNA molecules of substantial length at intermediate temperatures, the helix melts in more or less distinct steps; so long as the strands do not dissociate, all base pairs melt in contiguous groups of 50–400, called 'domains'. Within domains melting is completed over a small temperature interval. These properties are rationalized by a rigorous, complete two-state theory of the intramolecular melting transition (9, 10, 11). The relevant theory has been presented elsewhere (4, 12). Although improvements continue, particularly with respect to longer-range interactions and in allowing separation of entropic and enthalpic contributions, the present formulation remains generally useful. The theory and its derived calculations, which pertain only to unimolecular melting equilibria, apply only where the melted strands remain parts of a single molecule, either by virtue of a G,C clamp or a psoralen cross-link.

A large change in electrophoretic mobility on melting is observed only in a densely obstructed medium, such as a gel or a concentrated solution of a long chain linear polymer. When part of the molecule, at least 30 bp, remains helical, melting of 200 bp reduces the mobility to about one-tenth of that of the original helix, as compared to a much smaller reduction, perhaps to one-fourth, with full melting and separation of the strands. The low mobility of a partly melted molecule is attributable to its form as a three or four ended particle; the helical G,C-dense portion, the clamp, constitutes one end, and the unravelled strands the other. The molecule can migrate only when random changes in configuration permit two arms to move through the same pore (13).

The drop in mobility at the melting transition of end cross-linked molecules, as prepared by the psoralen method, occurs as the relatively stiff double helix changes to a double length flexible chain.

The theory has guided development of the technology from one offering a good chance of finding a mutation (14) to one with reasonable certainty that a sequence difference will be perceived (15). It continues to constitute an essential part of the experimental planning within the entire DGGE family. From the theory, it is clear that:

(a) Melting without strand dissociation is necessary, hence introduction of the G,C clamp and the psoralen cross-link.
(b) Thermal stability patterns can be altered and made more favourable by end-modification of fragments.
(c) Points of division for segmenting a long sequence into workable fragments can be specified rationally (16).

The theory is expected to continue to be important for further developments in mutation detection.

4. Overview of the procedures in searching for mutants

4.1 Defining segments for scrutiny

The first step in effective laboratory mutant detection within genomic DNA of known sequence is selection of segments containing putative mutants and planning modification of the thermal stability pattern of the segments. This step is carried out by computation of the distribution of thermal stability along the molecule, resulting in a pattern termed the meltmap. The computation obviates the need for lengthy trial and error laboratory effort. We review here the use of the family of *MELT* programs, in particular, *MELT94*, which provides easy computation of meltmaps. The meltmap is the tool for:

(a) Selecting favourable 5′ and 3′ ends for a segment to be scrutinized.
(b) Designing 'clamp' extensions to optimize detection.
(c) Specifying in detail the sequence interval within which sequence variants will be detected.
(d) Suggesting segment end modification to augment the interval.
(e) Anticipating the temperature–denaturant conditions needed for discrimination.

We describe also a program, *MUTRAV*, for estimating relative migration velocities at uniform denaturant concentration (as in CDGE, CDCE, etc.) and for estimating optimal conditions for separation of mutant and wild-type bands in a gradient.

In general, mutants will be sought in a genomic region for which the base sequence has already been determined. The region must be broken for analysis into segments of roughly 150–400 bp, usually with a small overlap to preclude detection failures at the ends of segments. The segments are of convenient length for PCR amplification. The length and precise positioning of the segments is inferred by computation of the meltmap of each proposed segment, as discussed in Section 6.2.5. The end-sequences should have favourable primer properties, as indicated by the usual criteria.

4.2 Sample preparation

Once an appropriate segment has been identified, PCR amplification is carried out using one primer carrying a 5' extension, unrelated to the genomic sequence, to serve as a clamp, both to suppress strand separation and to provide domain leveling. The extension can be either a G,C-dense sequence of at least 30 bp or a modified base carrying a tethered psoralen moiety. For fluorescent detection of band transit in the capillary method, one primer carries a covalently bound 5' fluorescent label.

It is desirable to convert each product of PCR amplification of genomic DNA to a heteroduplex by reassortment of strands with and without similarly amplified wild-type DNA. The separate PCR samples are mixed, heated, and cooled. For psoralen clamping, reassortment precedes irradiation.

4.3 Gradient and velocity separations

The standard denaturing gradient, the urea–formamide gradient in a polyacrylamide gel at 60 °C provides the basis for most applications. Complete details of the laboratory manipulations for simple gradient gels have been presented previously (17); expanded to include G,C clamps (18, 19); and slightly updated (20). Since they have not changed substantially, detailed repetition is unnecessary.

The amplified samples are applied to a gel in a thermostatted bath, or fed into a capillary maintained at a fixed temperature, or applied to a gel with a controlled temperature gradient. Close temperature regulation and uniformity in the bath are desirable. The field is applied. Neither running time nor closely controlled field strength are critical, because of the very low mobility of the molecules after partial melting. In a gradient gel the result consists in differences among a set of final gel positions reflecting retardation levels in the gradient, not mobilities.

In the absence of a gradient, the molecules migrate at characteristic mobilities, indicated and recorded either as the distance moved in a defined time or as the time needed to traverse the gel (or a polymer solution in a capillary).

Methodology for velocity separations in uniform denaturant has been presented (21). For methods in the less prevalent procedures, see the following: for capillary electrophoresis (22); thermal gradients (23, 24); and temperature ramp (25) and the D-Code Manual (Bio-Rad, Hercules, CA. The procedures for two-dimensional separations have been described (26).

4.4 Features of the gel patterns

The four band quadruplet will be seen in mobility separations as well as in gradient retardation. Substitutions consisting of conservative tranversions, AT$\leftrightarrow$TA or GC$\leftrightarrow$CG (27) and some single base deletions or insertions, may not result in separation of the mutant and wild-type homoduplex bands, and rare heteroduplex pairs are not resolved. However, heteroduplexes from any substitution mutation, which contain a mismatch, or from an insertion–deletion, which contain a loop, are always resolved from the homoduplex as a band at lower denaturant concentration in the gradient, or more slowly migrating in uniform denaturant. Very little separation within both the homoduplex pair and the heteroduplex pair, although the pairs are resolved, has been found among mutants of the human mitochondrial tRNA genes (28).

The presence of two sequence variations in the same domain will result in larger intervals between the homo- and heteroduplex bands, roughly the sum of the individual effects, if the sites are not too close. The interval between the homoduplex bands may be either larger or smaller than for a single sequence difference. We have seen one example in which the substitution of two adjacent bases retains homo–heteroduplex separation, but a smaller separation than a single substitution (Abrams, unpublished data).

Obviously, phenotypically undetectable polymorphisms will generate the same indications of a sequence difference, and discrimination will depend on independent biological information.

4.5 Discrimination of zygozygosity

If the source genome is heterozygous for a particular base change, heating and annealing to effect strand reassortment without addition of authentic wild-type DNA will generate the standard four band quadruplet. Heating and annealing of a mutant with added authentic wild-type DNA gives a quadruplet, regardless of the zygosity. Thus it is advisable to apply both the pure heating–annealing product of the test DNA and that heated–annealed with wild-type DNA to the gradient gel to distinguish homo- and heterozygotes. This consideration is particularly relevant where the phenotypic effect of a recessive gene depends on loss of heterozygosity, as with tumour suppressor genes.

Where the frequency of mutant alleles is high, as for thalassaemia in some Mediterranean regions, the recessive phenotype sometimes arises from different mutants in the parents. In this event, reassortment without addition of wild-type DNA shows a quadruplet, and reassortment in the presence of wild-type DNA yields nine bands, not necessarily all resolved.

Two variants from different sources in the same domain may exhibit indistinguishable quadruplets even though the sites of the variants differ. Whether they are identical or not is directly resolved by melting–annealing mixed PCR products. If identical, and both are homozygous, there is one displaced band; if identical and one or both are heterozygous, there is a clear quadruplet; if they differ and one or both is heterozygous, nine bands are expected.

4.6 Comments

Clamps are easily incorporated during sample preparation as an extension of one of the PCR primers. The electrophoretic procedure closely resembles ordinary gel electrophoresis with a little additional intricacy. The gel methods require polyacrylamide, rather than agarose to withstand denaturing conditions; 10–30 minutes must be spent in pouring a gradient; and the gel must be immersed in a thermostatted water-bath. The mobility method, which eliminates need for a gradient, is satisfactory for repetitive screening of the same portion of a gene sequence. Gradient mixing in preparing a gel is also unnecessary in thermal devices, discussed below, which give essentially the same results as urea–formamide gradients. It has been noted that the quadruplet is so conspicuous that broad, low resolution gradients are efficient in allowing fragments of diverse T_m to be analysed in one gel (29, 26). Abrams *et al.* (27) and others have loaded several fragments of differing T_m into the same lane.

5. Use of the psoralen cross-link as a clamp

G,C clamping and psoralen clamping appear to be interchangeable, but the psoralen method offers some benefits. Psoralen cross-linking was introduced as a an alternative to the G,C clamp (2, 30), following a report of site-specific cross-linking in triple helices (31). For a sequence consisting of a high T_m domain, the G,C clamp may not provide a sufficient margin of stability to permit distinct retardation prior to strand dissociation, but the covalent psoralen link does not permit strand separation. Where a supplier provides a large number of oligos, the incremental cost of the psoralen-modified 5' termination may be appreciably less than a set of 30 bp G,C extensions.

There are, however, two minor disadvantages to the psoralen method.

(a) The UV cross-linking is never 100% complete, and a small amount of the uncross-linked fragment may add an extraneous band to the gel pattern (see *Figure 1*, lane 4). The band is easily distinguished; it is rarely focused, it is not retarded, and it is likely to disappear in a longer run. Cross-linking may be less complete in small fragments, 100 bp or fewer.

(b) Cloning after recovery from the gel requires removal of the cross-linked end by a restriction endonuclease. If the sequence lacks an obvious 5' site, a site can be included in the primer. An overhanging 5' end after restriction can be filled to match the 3' blunt end.

Psoralen addition slightly decreases the electrophoretic mobility of the double-stranded fragment in agarose gels, indicating an apparent length a little larger than expected.

Optimal cross-linking by photoactivation of psoralen requires pyrimidines, preferably thymidines, on opposing strands in adjacent pairs bracketing the psoralen intercalation site (32). The psoralenylated oligonucleotide is easily prepared by conventional solid phase synthesis, using a phosphoramidite of 5 (ω-hexyloxy)-psoralen after oligomer chain elongation. The deprotection is done by treatment with 0.4 N NaOH, and finally, the product is purified by ion exchange reverse-phase HPLC. Oligos carrying 5′ psoralen are available from various suppliers, and the phosphoramidite is available from Appligene and Glen Research. From Glen it is identified as 2-[4′-(hydroxymethyl)-4,5′,8-trimethylpsoralen]-hexyl-1-*O*-[(2-cyanoethyl)-*N*,*N*-diisopropyl)]-phosphoramidite.

Psoralen cross-linking has been used effectively for mutation detection in 27 exons of the cystic fibrosis transmembrane regulator gene, (33), in β-globin for thalassaemia (2), in multiple endocrine neoplasia and sporadic pheochromocytoma (34), and in human mitochondrial tRNA genes (27).

5.1 The psoralen protocol

5.1.1 Amplification

The 5′ end of the primer for the end of the segment at which a clamp is optimal, as inferred from the meltmap, must carry an appropriately linked psoralen moeity at the end adjacent to T or A, preceding the genomic sequence (2). The optimal primer sequence may be 5′(Pso)pTpApnpnp.....3′, given the preference of psoralen for binding between TpA and ApT pairs (35). This 5′ modification is offered by some suppliers and the necessary phosphoramidite can be purchased for in-house primer synthesis. Otherwise, PCR amplification is conventional; the 5′ psoralen has negligible effect on PCR conditions.

5.1.2 Cross-linking

Load a sample of the PCR product into a flat-bottomed microtitre plate; support the plate under a 365 nm UV source. At 0.5 cm from an 8 W portable UV lamp, 15 min is sufficient. If the volume is small, minimize evaporation by cross-linking at a lower temperature, 4–10 °C; the yield is not altered. The samples can be loaded into a gel immediately, or they can be stored at 4 °C or −20 °C until needed.

5.1.3 Evaluation of cross-linking

The efficacy of cross-linking can be evaluated by the electrophoretic analysis in a polyacrylamide denaturing gel. The bond formed by the psoralen at one end of the DNA fragments results in single-strand molecule of twice the double-stranded length, giving a retarded band. Optimal cross-linking time at room temperature was estimated to be 15 min, with a yield of about 80%. This reaction does not depend on the composition of the PCR buffer. The integrity of DNA after one hour of irradiation is indicated by the continued formation of a single, compact band in fully denaturing polyacrylamide.

G,C clamped and psoralen clamped sequences show similar melting behaviour. The unmelted psoralen clamped fragment has a somewhat higher electrophoretic velocity, presumably because it lacks the 30 bp increment of the G,C tail. A meltmap calculated for a G,C clamp remains appropriate for the psoralen clamp.

6. Computational tools

Calculation of the meltmap precedes all laboratory procedures to establish favourable segment end-points, to select the end for a clamp or cross-link, and to test the benefit of end-modification. The calculations can be carried out in less than a minute on a PC from a file containing the base sequence of the molecule, using a program, such as *MELT94*, based on statistical mechanical theory. Somewhat similar programs are also available from other sources (see below).

6.1 What is a meltmap?

The meltmap shows the equilibrium temperature, T_m, along the sequence at which each base pair in the molecule has equal probability for the helical or melted states, so long as the strands remain associated. It describes the progress of melting equilibrium within a partly melted molecule as the temperature rises. At temperatures below the curve, a base pair is most probably in the helical conformation; at temperatures above the curve, the pair will most probably be melted, non-helical. It is derived by means of the Poland–Fixman algorithms directly from the base sequence.

Figure 2A presents a smoothed plot of local variation in the ratio of A,T to G,C base pairs in a portion of the human testes determining gene, *sry*. It is obvious that this is a jittery function, and that if base pairs melted individually or as clusters defined by some window of averaging, the molecule would be a ragged, bubbly string at intermediate temperatures. *Figure 2B* shows the meltmap of the sequence calculated using *MELT94* (20 sec on a 133 mHz Pentium). There is clear local coherence of the T_m, but it is not uniform across the fragment. *Figures 2C* and *D* are meltmaps of the same sequence which has been extended by 30 G,C base pairs at the 3' or the 5' end. These ends remain helical at temperatures sufficient to melt the genomic segment; thus they serve as clamps. *MELT94* includes a simple command for addition of a 30 bp clamp at either end, as specified. No significant difference is found for most segments with longer clamps.

The benefit of 5' clamp extension is clear in the map in *Figure 2D*; the fragment melts as a single, co-operative unit, a domain, from bp 62 to bp 265.

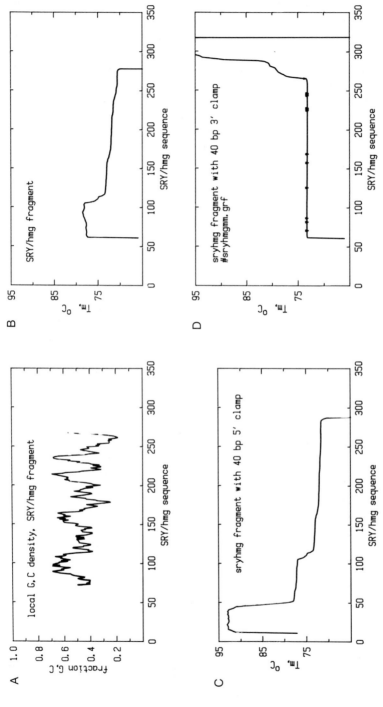

Figure 2. Local G,C density and differently clamped meltmaps of the *hmg* fragment of the human *sry* gene. (A) The local G,C density, represented by polynomial smoothing with G,C pairs = 1.0, A,T pairs = 0.0. (B) Meltmap calculated by *MELT94* for the fragment. (C) Meltmap calculated with a 30 bp G,C clamp at the 5' end. (D) Meltmap calculated with a 30 bp G,C clamp at the 3' end. Sites of known mutations are marked.

The importance of the meltmap for mutation searching derives from its specification of the span of the sequence in which a sequence variation is readily detected. Any mismatch within the level span of the domain with the lowest T_m can be expected to raise the retardation level of the fragment to a lower denaturant concentration; it decreases the stability of the helix. Almost any base substitution would shift the retardation level of a homoduplex up or down, relative to wild-type. The amount of shift depends both on the specific substitution and the neighbouring pairs. *Figure 2* indicates the positions of known mutants; all are detectable as shifts in the retardation level of the molecule, relative to normal *sry*. Correspondingly, any departure from this wild-type sequence within the span of that domain will be revealed as a change in migration velocity in uniform denaturant if the denaturant is close to the transition concentration.

The meltmap of the β-globin fragment with a 5' clamp generating the bands in *Figure 1* is shown in *Figure 3*. A meltmap is usually so slightly altered by a single base change that a plot based on the modified sequence is not informative unless the change is within a few bases of a boundary to a slightly more stable domain. The splice site mutation, G to A at 359, the domain boundary, in *Figure 3* is an example; the map is plotted from the substituted sequence. This fragment is useful in permitting mutations to be detected in three domains; the second lowest in *Figure 3* becomes the lowest after cleavage with *Mbo*II at 435–6, and the next becomes the lowest after cleavage at 267–8 with *Nla*III.

As the equivalent temperature rises, the equilibrium distribution between

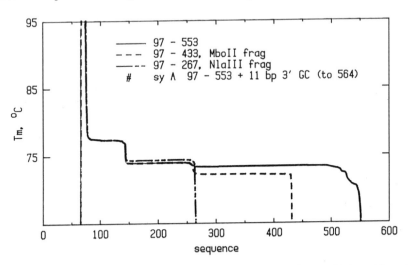

Figure 3. Meltmap of the normal β-globin fragment which, in a heteroduplex with mutant fragments, produced the patterns in *Figure 1*. The meltmaps of the *Mbo*II and *Nla*III truncated fragments are included. All are calculated with a 5' 30 bp clamp simulating the cross-link.

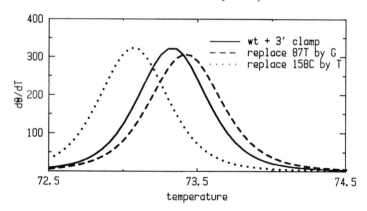

Figure 4. The melting transition for the principal domain of the 3' clamped *sry/hmg* fragment, also for the same molecule with a T to G substitution at position 87 and with a substitution of T for C at 158. The transition is shown as the derivative of the calculated probability of non-helicity, summed over all base pairs. Values of the function are given by *MELT94* for any sequence as calc/<filename>.dth.

the helical and melted states shifts. The temperature interval for the transition is shown on the computer display as a pair of curves bracketing the meltmap, one indicating the temperature for a 20–80% distribution and the other for an 80–20% distribution. The calculated first derivative of the amount of melting as a function of temperature is shown in *Figure 4*, representing the total of all bases in the flat domain of the 3' clamped *sry* fragment and two mutant fragments, T→G at position 87 and C→T at 158.

The plot shows the narrowness of the melting transition for a long, flat domain and the shift due to substitutions. Bands in the denaturing gradient, whose width is initially defined by loading conditions, become sharply resolved at retardation level because of the reduced mobility; focusing does not occur in the absence of a gradient.

6.2 Meltmap protocol

6.2.1 Acquiring the *MELT* program

The most direct way to initiate calculation procedures is to download the program, *MELT94*, from the World Wide Web into an IBM compatible computer using the DOS operating system, which is available in Windows. Once the program is installed, maps can be calculated without further dependence on the Web.

Note: DOS/Windows remains the operating system of choice among personal computers because of the availability of the programs described here and others that may be expected. Low cost computers are satisfactory; old models are OK if they contain a math coprecessor. Programming language limitations render the Mac useful only with compiled, proprietary meltmap programs.

MELT87, which lacks the user-friendly features of *MELT94*, is available from the Web as FORTRAN source code, and it can be compiled for execution on other computers.

The Web address is `http://web.mit.edu/osp/www/melt.html`. On entry to this Web site, the display will present the graph and text shown in *Figure 5*. The plot shows the meltmap of an unclamped fragment of the human β-globin sequence lying between two *Hae*III restriction sites.

Bring the cursor arrow to the words, 'Download the program here', and click. A set of compressed files will be installed into the DOS directory. At the DOS prompt, type the command, `pkunzip  -d  meltmap.zip` [enter]. Your directory will contain the executable program, `MELT94.exe`, together with the subdirectories, `seq`, `set`, `lib`, and `calc`, and additional (hardly necessary) advice in `MELT.txt`. These downloading instructions are presented on the Web page.

6.2.2 Computer operations

Concise instructions are included in the program, and additional advice is provided in `MELT.txt`, which should be printed for reference. The program is invoked by typing `MELT94` [enter]; after the title page it presents a menu requesting entries for details of the proposed calculation, as shown in *Figure 5*.

To set-up the calculation for a particular sequence, type the values appropriate for each item, one-by-one, by typing first, the item number [enter], followed by the value or the name [enter]. The new value will appear on the menu after each entry. Watch the top or bottom of the screen after each entry as you go; you will be advised of errors or out-of-bounds values.

The sequence interval between items 1 and 2 cannot be longer than 1000 bp without a clamp, or 970 bp with. Since the effect of a psoralen cross-link is correctly simulated by the G,C extension, it is not specified explicitly; the cross-link is represented by a clamp, and the 970 bp limit applies.

Item 3, the temperature from which the calculation begins must be at least as low as the lowest T_m in the molecule. Usually 60 °C is satisfactory, and no T_m in the resulting meltmap will be below 60 °C, but rare A,T-dense regions may be cut off and require 55 °C or 50 °C. A value of 0.2 °C for item 4 gives a fine-grained result, but a larger value is faster for a rough map. Item 5 must exceed the highest T_m in the molecule; if not, an error message will appear.

The name of the source sequence (item 6) must precisely match the name of a file in the `seq` subdirectory. For item 7, give the results a unique, recognizable name, including an indication of the clamp position, if any. Do not use more than eight characters total, no spaces. The same name is never to be used again if the results are to be saved; otherwise the next calculation will overwrite the first.

Note the opportunity at item 8 to extend the sequence at either end with an arbitrary 30 bp C,G segment simply by typing 3′ or 5′ instead of N; type N,

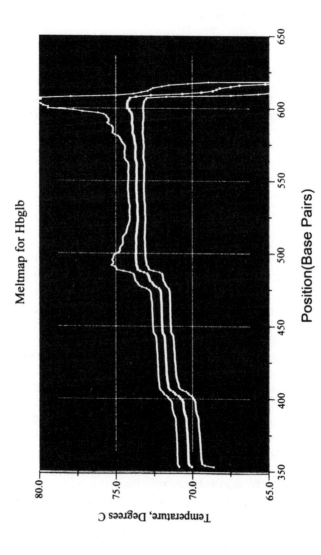

Figure 5. *MELT94*, as it appears at the Web site: http://web.mit.edu/osp/www/melt.html.

THE CURRENT SETUP PARAMETERS FOR CALCULATIONS ARE:

1. Position of first basepair in fragment __
2. Position of last basepair in fragment____
3. Lower temperature bound for calculations_
4. Temperature increase between iterations__
5. Upper temperature bound for calculations_
6. Name of file containing source sequence__
7. Output filename, (no extension)_____
8. Clamp attachment_____

TO REVISE THE SETUP SPECIFICATIONS, TYPE THE ITEM NUMBER
AND PRESS [ENTER]; THEN TYPE THE NEW VALUE OF THE PARAMETER
AND [ENTER]. REPEAT FOR EACH OF THE OTHER PARAMETERS.
WHEN THE PARAMETERS HAVE BEEN ADJUSTED, TYPE S [ENTER]
TO START THE CALCULATIONS.'
    If it should be necessary to abort the calculation
    at any time, press <CTRL> and C simultaneously.

Figure 6. The first menu in execution of *MELT94.*

if no clamp is to be added. Do the same for a 5′ or 3′ psoralen; its effect is simulated properly by a G,C clamp.

Since the program saves all parameters in the menu when it ends, the calculation can be repeated rapidly with a new clamp specification (or any other parameter) by entering melt94 [enter] again and revising both items 7 and 8 (or other values). When the menu entries are correct, press [enter] to initiate calculation; progress is indicated at the bottom of the display. When it is finished and [enter] is pressed, the highest and lowest calculated T_m values are shown.

To calculate properties of sequence alteration, other than 30 bp clamp extension, copy the sequence source file to a new file with a modified name, and insert the alterations with a file editor.

6.2.3 A plot of the meltmap on the display

The plot menu is presented, requesting appropriate sequence and temperature ranges for the plot coordinates—the length of the segment and the temperature limits. It is unnecessary that these be optimal on the first try; they can easily be readjusted after the first plot is presented, and the temperature limits can be broad for a complete map or very narrow for a better view of details. Unless the computer is linked to an appropriate printer (see below), item 4 should remain n.

If the reader specifies the β-globin sequence, hbglb, and segment limits, 348 and 720, with no clamp, at the start of the *MELT* program, the plot shown on the display after typing p [enter] will reproduce *Figure 6*. Recalculation with a clamp will be instructive. Note that this segment also includes the mutants represented in *Figure 1*.

In addition to the meltmap, which is defined as the temperature of 50–50 equilibrium between helix and melting for each base pair, the plot also includes contours for 20–80% and 80–20% equilibria. The screen is cleared of the plot by pressing [enter] if no printing was specified (n at item 4). For hard copy, revise the plot title, item 7. The menu will be presented again with its previous entries for revision. The program may be terminated at any time with [ctrl].

In MS-DOS running under Windows 95, the full-page display may shrink on exiting from the plot. Click on the full-page icon in the toolbar to return to the program.

A copy of the screen display can be printed on recent models of Hewlett–Packard laserjet printers and some others if item 4 of the plot menu is set to Y. Transfer to the printer may be slow. If the page is not delivered on restoring the display and ending the program, invoke a formfeed. If item 5 was set to Y, instead of N, and the plot is not printed or no printer is connected, the system can be restored to action only by rebooting; press three keys together—alt, ctrl, and delete.

6.2.4 Numerical tables of the results

On exiting from *MELT94* after clearing the plot, if the plot is not to be repeated, type n [enter]; the screen lists the files containing the numerical results deposited in the calc subdirectory. The content of each of the files is shown on the list. The meltmap file with the suffix, *.mm5, is in a two column format, which can be copied into most graphics programs for a hard copy plot. However, a printout will require many pages, and a compact, printable tabulation of the meltmap is contained in *.5mm. The *.tdk file lists temperatures at which an approximate calculation of the dissociation constant for strand separation reaches values of 10^{-3} to 10^{-7} M^{-1}. It is unlikely that useful bands will be seen in a gradient gel if the lowest temperature listed in *.tdk is below the T_m of the domain scrutinized. The suffix, *.mus, designates the file containing the initial set-up parameters. The suffix *.muf designates the file conaining the sequence of the fragment mapped, as extracted from the source sequence, with its clamp, if any.

The melting profile is indicated by the derivative of the number of bases melted as a function of temperature, tabulated in *.dth.

6.2.5 Practical application of the calculations

Most human mutation searching is focused on exons and immediately adjacent sites, which can be contained within segments of length 100–400 bp, suitable for DGGE. Long exons must be divided into smaller segments. Since there is no simple algorithm for selecting appropriate segments, some guesswork and trial and error calculations are needed to pick sequence limits and the G,C or psoralen clamp positions that generate satisfactory domains. The calculation merely requires entry of the selected end-points and designation of the end to be clamped, and planning can be rapid. The long-range effects of clamps in flattening meltmaps is surprising, and the choice of a 5' or 3' end makes a major difference, as has been shown in *Figure 2A–D*.

For exons of about 50–400 bp, begin calculating trial segments by selecting end-points a few base pairs into the introns, away from the exon boundary, to include splice sites. Longer extension into an intron is likely to introduce an extraneous low melting section. Very short sequences are unlikely to give good domains or sharp retardation. Dips and end droops can be corrected with insertion of extra bases in the primer. For regions other than exons, preliminary large scale meltmapping (below) is helpful. One end can be chosen in a stable (high melting) region, the other ad lib, but the meltmap of an excised fragment may depart widely from a that of a longer sequence. Avoid fragment lengths much longer than the least stable (lowest melting) domain. Very long lowest melting domains, greater than 400 bp, are rarely encountered; it may be better to divide the sequence, since band intervals in the quadruplets decreases with increasing domain length.

If the map of the selected segment does not immediately transform into a plateau and peak pattern with the addition of a clamp at one end or the other, trial and error testing of new end-positions will be needed.

If it is of interest to examine a long continuous sequence, 1 kb segments can be chosen with at least 200 bp overlap for concatenation into a continuous long map, with inspection for jogs to insure adequacy of the ovelap length.

Since irregular contours or sharp changes may obscure limits of the domain, particularly at an end or close to the clamp, one should verify the span of the level domain more precisely from the numerical files reported on ending the program, calc\<name>.mm5 and calc\<name>.5mm.

Mutations may also be detected in a part of the sequence a few base pairs away from the lowest level domain if the temperature increment is small. Calculation by *MUTRAV* is recommended or recalculation of the map with a hypothetical substitution at a boundary between domains differing by less than about 3 °C in T_m.

Segment limits that result in a low melting domain sandwiched between more stable domains should be revised, if possible. Mutants within that domain can be discerned, but the bands are broadened, making gradient shifts less conspicuous.

6.3 Predicting electrophoretic separations

The program, *MUTRAV*, simulates the dependence of the rate of electrophoretic migration in a gel on the extent of melting so long as a sufficient part of the molecule remains helical. It includes provision for specifying sequence substitutions without altering the source file and provision for inserting an arbitrary stability perturbation to simulate mismatches. *MUTRAV* presents results in three ways:

(a) The electophoretic velocity as a function of temperature.

(b) The depth in the gradient of a molecule as a function of the time of electrophoresis.

(c) The difference in gradient depth between the wild-type and mutant homoduplexes as a function of the time of electrophoresis.

These are presented both as graphs on the display and as numerical tables that can serve as input to further calculation or to a hard copy plotting program. The velocity plot displays the change in electrophoretic mobility determined by the extent of melting. It is directly applicable to separations in a uniform denaturant concentration at constant temperature, discussed below. The program allows variation of the clamp length, it displays the expected effect of each clamp length, and it incorporates convenient means for introducing sequence substitutions.

The gradient difference calculation (c), is most useful; it corresponds to the informative feature of the gel pattern of DGGE with homoduplexes. The differences can be positive or negative, depending on the substitution. Display of the time dependence of the difference between wild-type and mutant bands is particularly valuable when the map of the sequence is not entirely flat; in that event, mutants at some positions reach maximum bandshift at differing running times. The program accepts multiple substitutions (but the results may be grossly incorrect if they are adjacent or close).

6.3.1 Migration velocities in uniform denaturant

The velocity vesus temperature calculation, is the counterpart of separations at constant temperature in a uniform denaturant concentration, as in perpendicular gradients, in the capillary system, and in CDGE. The calculation is useful in determining the narrow temperature interval within which substantial mobility differences due to a substitution (or a mismatch, as above) can be demonstrated.

Figure 7 shows the calculated ratio of mutant to wild-type homoduplex mobilities for a number of substitutions in a p53 fragment, which melts as a single domain because of a clamp. It can be seen that the equivalent temperature (thermometer temperature plus denaturant, as above) must be selected within narrow limits for separations to be detected. However, at the optimum temperature, very large separations can be expected, up to threefold differences in band movement for this molecule. The stability differences determined by the nature and context of the substitution are clearly resolved. Large differences in mobility, as predicted here, are best followed in capillary

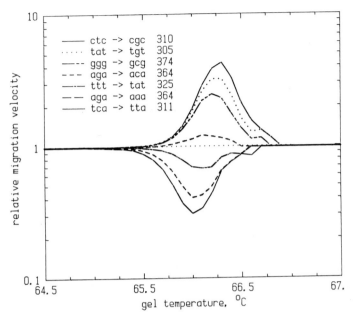

Figure 7. The ratio of gel electrophoretic mobilities of mutant p53 fragments to the mobility of the normal fragment, as a function of temperature, both clamped.

electrophoresis, where the bands are sharp and migration time is recorded by the instrument.

These calculations are only approximate and in relative units, in that they depend on a rough theoretical model of electrophoresis and an approximate relation between temperature and denaturant concentration. They do not include correction for denaturant viscosity. Nevertheless they are reasonably reliable in reflecting the effects of sequence differences. The two parameters describing equivalence between calculated temperature and denaturant concentration differ with details of the gel, and various relations have been published. The differences at middle denaturant concentrations are small among various proposed parameter pairs, such as:

$$T_{calc} = 49.0 + (\% \text{ denaturant})/2.0$$
$$T_{calc} = 57.0 + (\% \text{ denaturant})/3.2$$

6.3.2 Band separations in a gradient

The difference in retardation levels for three substitutions in the clamped human β-globin fragment (*Figures 1* and *3*) are shown in *Figure 8* as a function of the running time of the gel, calculated by *MUTRAV*. The plot shows that variants in the lower domain (359 and 469) reach and mantain large separations at about the same time, 6–12 h. A variant in the second level domain (not included in the figure) can be detected, since the T_m is only a little higher, but additional time is needed, and the first domain separations decline. Similar calculated and measured band separations are compared in ref. 16; one mutant that appears quickly might be missed (and was, initially when the running time was too long).

6.3.4 Band separations for heteroduplexes

Reassociation of the strands of wild-type and mutant molecules yields four different products, reconstituting the two original types and making two different molecules, each containing a different mismatch or a different looped-out base. Typically, all four are resolved in the gradient as a quadruplet, but in some sequences either the homo- or the heteroduplexes may not be separated. Because the destabilization at a mismatch site in a heteroduplex varies substantially with the sequence context (104 varieties of the triplex of the mismatch and nearest neighbours) and these effects have not been calibrated, the program cannot offer a satisfactory prediction of mismatch effects. However, most mismatches give band shifts at least as large as a G to A substitution; and arbitrary insertion of pairs of such substitutions permits a rough estimate of heteroduplex effects. Alternatively, an arbitrary value for a destabilization parameter can be introduced at the site of mutation, representing the lowered local stability in the vicinity of a mismatch. It is entered at the bottom of the *MUTRAV* menu. Appropriate values for this parameter, estimated by fitting *MUTRAV* calculations to gel data, range from about –30°C to –70°C.

6.4 Computer operations for *MUTRAV*

Invoke the Web site: `http://web.mit.edu/biology/ lerman/ mutrav.html`

Download `mutrav.fls`, apply `pkunzip -d`, (same as the procedure for *MELT94*). Your directory will contain `MUTRAV.exe`, four subdirectories, of which `set`, `seq`, `lib`, and `calc`, which correspond to the same usage as in *MELT94*, and `MUTRAV.doc`, which contains complete, user-friendly details on running the program. It is set-up like *MELT94*, but it includes means for introducing sequence substitutions or alterations in helix stability at specified sites. Full instructions are contained in the file, `MUTRAV.txt`.

The calculations result in plots on the display and the numerical files:

(a) `calc\<name>.mus`—the specifications from the first menu.

(b) `calc\<name>.muf`—sequence of the fragment, as clamped.

(c) `calc\<name>.gmu`—relative mobility versus equivalent temperature, two column format, for plotting.

(d) `calc\<name>.gy`—gel position (cm from top) versus time, two column format, for plotting.

(e) `calc\<name>.gyd`—separation of prototype and mutant bands (cm) versus time, two column format.

(f) `calc\<name>.mu`—same as `.gmu`, compact format, printable.

(g) `calc\<name>.ylv`—same as `.gy`, compact format, printable.

(h) `calc\<name>.ydf`—same as `.gyd`, compact format, printable.

7. Other members of the DGGE family

7.1 Gel separations in a uniform, partially denaturing environment

Large separations of wild-type and mutant bands can result from careful selection of a single denaturant concentration (6, 36) such that partly melted molecules migrate at a constant velocity determined by the extent of unravelling. By trial and error refinement of the calculated temperature and gradient concentration homoduplex and heteroduplex bands are well resolved (37). The calculation of migration velocities using *MUTRAV* for both mutant and reference sequences indicates the separations that can be expected, together with temperature–denaturant conditions necessary for resolution. Bands differing by a factor of two to four in migration velocity, as in *Figure 7*, will be far apart in the gel or in exit time in capillary electrophoresis. However, this benefit is gained at the expense of requiring precise advance determination of the optimal temperature–denaturant conditions and loss of the band sharpening effect of gradient retardation. The method is particularly useful for examining a large number of corresponding segments from diverse individuals (21). Meltmap and clamp calculations have the same role as in gradient operations.

7.2 Capillary electrophoresis

Separation of wild and mutant-types in a capillary follows the same principles, but application of a field strength an order of magnitude higher than in slab gels reduces the time required correspondingly—to less than 30 minutes. A dense solution of high molecular weight linear polyacrylamide is used instead of a gel; the lack of polymer cross-links is no significance. Conventional capillary electrophoretic instrumentation enables automatic recording of the migration velocity. This system has been applied for determination of substitutions in *ras* (38) and other genes. The procedure is described in detail in ref. 21. A second cycle of amplification following intitial separation has permitted detection of a mutant sequence present as only 10^{-6} of the majority component (3).

7.3 The thermal gradient

Operations and observations with a thermal gradient (39) are the same as in conventional DGGE but simpler in that there is a uniform, one solution gel, and the gradient is established by setting temperature controls for the top and bottom of the gel. Although the added convenience has given useful results

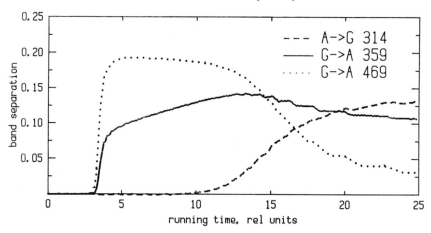

Figure 8. Calculated separation between homoduplex mutant and wild-type bands of clamped fragments of β-globin (*Figure 3*) as a function of running time of the gel. Two of the fragments correspond to lanes 2 and 5 in *Figure 1*.

(40–43), general use of the technique waits for introduction of a reliable, low cost instrument.

7.4 The temperature ramp

Despite the terminology, TTGE, TSGE, this method (44, 45) does not depend on a gradient; it is a variant of velocity separations in uniform denaturant, improved by a temperature ramp, a spatially uniform temperature increasing with time, that makes close, *a priori* temperature selection unnecessary. The results are similar to those in a denaturing gradient, but focusing from a drop in velocity on retardation does not obtain, and careful adjustment of conditions is essential. Results have been reported for mutants in the CFTR gene in capillary electrophoresis (25) and in the prion protein gene (46) in slab gels. The capillary separations were carried out in an 8% solution of a linear polymer similar to but more hydrophilic than acrylamide, with 6 M urea, a field strength of 100 V/cm, and a temperature ramp from 45°C to 49°C over about 20 min.

7.5 2D length and gradient separations

Vijg and collaborators have made extensive use of separations of high multiplicity, mixed PCR products, and restriction digest, first by length and then in the orthogonal direction by DGGE (26, 47, 48), similar to ref. 5. Although a fairly broad gradient is required, the characteristic set of three or four bands from a substitution and mismatch in reassorted strands is easily discerned. This technique provides a high resolution display of a large number of VNTRs (49, 50), enabling clear discrimination of polymorphisms, and it can permit survey of many fragments, covering a long sequence, in a single gel. Ref. 51 reports products of 25-fold multiplex PCR of the RB1 gene resolved in one gel, covering 18 exons, revealing eight variants not previously recognized and 25 others. A detailed procedure is given in ref. 26.

8. End notes

Other versions of melting programs have been published (52, 53), and proprietary versions, generally derived from *MELT87* (12) can be purchased.

Principles and calculations on partial melting are thought to apply also to HPLC detection of heteroduplexes (54). Since a band in a gel constitutes an open system, that is components, reaction products, or interacting molecules are free to move into or out of the band, chemical equilibrium is only temporary. The higher mobility of dissociated strands relative to partly melted molecules promotes gradual dissipation of bands, depending not on equilibrium but on the rate of strand separation. The comparatively low dissociation rate at their retardation temperature of fragments containing part of a CpG island together with a lower melting domain has been used to select island sequences (55).

Acknowledgements

We are grateful for helpful discussion with Lisa Steiner. This work was supported by a grant from the National Institutes of Health and INSERM.

References

1. Guldberg, P., Henriksen, K., and Guttler, F. (1993). *Genomics*, **17**, 141.
2. Fernandez, E., Bienvenu, T., Desclaux-Arramond, F., Beldjord, K., Kaplan, J., and Beldjord, C. (1993). *PCR Methods Appl.*, **3**, 122.
3. Thilly, W., Khrapko, K., Cha, R., Hu, G., and Andre, P. (1994). *Nucleic Acid Res. Mol. Biol.*, **9**, 285.
4. Lerman, L. S., Fischer, S. G., Hurley, I., Silverstein, K., and Lumelsky, N. (1984). *Annu. Rev. Biophys. Bioeng.*, **13**, 399.
5. Fischer, S. G. and Lerman, L. S. (1979). *Cell*, **68**, 183.
6. Fischer, S. G. and Lerman, L. S. (1983). *Proc. Natl. Acad. Sci. USA*, **80**, 1579.
7. Myers, R. M., Fischer, S. G., Maniatis, T., and Lerman, L. S. (1985). *Nucleic Acids Res.*, **13**, 3111.
8. Sheffield, V. C., Cox, D. R., Lerman, L. S., and Myers, R. M. (1989). *Proc. Natl. Acad. Sci. USA*, **86**, 232.
9. Wartell, R. M. and Montroll, E. W. (1972). In *Advances in chemical physics* (ed. I. Prigogine and S. A. Rice), Vol. 22, pp. 129–203. Wiley-Interscience, New York.
10. Poland, D. (1974). *Biopolymers*, **13**, 1859.
11. Fixman, M. and Friere, J. J. (1977). *Biopolymers*, **16**, 2693.

12. Lerman, L. S. and Silverstein, K. J. (1987). In *Methods in enzymology* (ed. R. Wu, L. Grossman, and K. Modave), Vol. 155, p. 482–501. Academic Press, San Diego.
13. deGennes, J. G. (1979). *Scaling concepts in polymer physics*, Cornell University Press, Ithaca, New York.
14. Lerman, L. S., Silverstein, K., and Grinfield, E. (1986). *Cold Spring Harbor Symp. Quant. Biol.*, **LI**, 285.
15. Myers, R. M., Fischer, S. G., Lerman, L. S., and Maniatis, T. (1985). *Nucleic Acids Res.*, **13**, 3131.
16. Abrams, E. S. Murdaugh, S. E., and Lerman, L. S. (1990). *Genomics*, **7**, 463.
17. Fischer, S. G. and Lerman, L. S. (1979). In *Methods in enzymology* (ed. R. Wu, L. Grossman), Vol. 68, p. 183–191. Academic Press.
18. Myers, R. M., Maniatis, T., and Lerman, L. S. (1987). In *Methods in enzymology* (ed. R. Wu, L. Grossman, and K. Modave), Vol. 155, p. 501–27. Academic Press, London.
19. Abrams, E. S. and Stanton, V. P. Jr. (1992). In *Methods in enzymology* (ed. D. M. W. Lilley and J. E. Dahlberg), Vol. 212, p. 71–104. Academic Press, San Diego.
20. Murdaugh, S. E. and Lerman, L. S. (1996). In *Laboratory protocols for mutation detection* (ed. U. Landegren), p. 33–37. Oxford University Press, New York.
21. Borresen, A. (1996). In *Technologies for detection of DNA damage and mutations* (ed. G. Pfeifer), p. 267–79. Plenum Press, New York.
22. Hanekamp, J. S., Andre, P., Coller, H. A., Li, X. C., Thilly, W. G., and Khrapko, K. (1996). In *Laboratory protocols for mutation detection* (ed. U. Landegren), p. 38–41. Oxford University Press, New York.
23. Riesner, D., Steger, G., Zimmat, R., *et al.* (1989). *Electrophoresis*, **10**, 377.
24. Wartell, R. M. Hosseini, S. H., and Moran, C. P. Jr. (1990). *Nucleic Acids Res.*, **18**, 2699.
25. Gelfi, C., Righetti, P., Cremonesi, L., and Ferrari, M. (1994). *Electrophoresis*, **15**, 1506.
26. Li, D., Van Orsouw, N., Huang, C., and Vijg, J. (1996). In *Technologies for detection of DNA damage and mutations* (ed. G. Pfeifer), p. 291–305. Plenum Press, New York.
27. Abrams, E. S., Murdaugh, S. E., and Lerman, L. S. (1993). *Genes, Chromosomes Cancer*, **6**, 73.
28. Michikawa, Y., Hofhaus, G., Lerman, L. S., and Attardi, G. (1997). *Nucleic Acids Res.*, **25**, 2455.
29. Guldberg, P. and Guttler, F. (1994). *Nucleic Acids Res.*, **11**, 880.
30. Costes, B., Girodon, E., Ghanem, N., *et al.* (1993). *Hum. Mol. Gen.*, **2**, 393.
31. Takasugi, M., Guendouz, A., Chassignol, M., *et al.* (1991). *Proc. Natl. Acad. Sci. USA*, **88**, 5602.
32. Cimino, G. D., Gamper, H. G., Isaacs, S. T., and Hearst, J. E. (1985). *Annu. Rev. Biochem.*, **54**, 1151.
33. Bienvenu, T., Cazeneuve, C., Kaplan, J., and Beldjord, C. (1995). *Hum. Mutat.*, **6**, 23.
34. Beldjord, C., Desclaux-Arramond, R., Sanson, M., *et al.* (1995). *J. Clin. Endocrinol. Metab.*, **80**, 2063.
35. Gamper, H., Piette, J., and Hearst, J. E. (1984). *Photochem. Photobiol.*, **40**, 29.
36. Lerman, L. S., Fischer, S., and Lumelsky, N. (1983). In *Recombinant DNA and medical genetics*, p. 157–86. Academic Press.

37. Hovig. E., Smith-Sorensen, B., Brogger, A., and Borresen, A. L. (1991). *Mutat. Res.*, **262**, 63.
38. Kumar, R., Hanekamp, J., and Louhelainen, J., (1995). *Carcinogenesis*, **16**, 2667,
39. Thatcher, D. R. and Hodson, R. (1981). *Biochem. J.*, **197**, 105.
40. Rosenbaum, V. and Riesner, D. (1987). *Biophys. Chem.*, **26**, 235.
41. Ke, S. H., Kelly, P. J., Wartell, R. M., Hunter, S., and Varma, V. A. (1993). *Electrophoresis*, **14**, 561.
42. Wartell, R. M. and Ke, S. H. (1995). *Biochemistry*, **14**, 4593.
43. Shiraishi, M., Murdaugh, S. E., and Lerman, L. S., in preparation.
44. Yoshino, K., Nishigaki, K., and Husimi, Y. (1991). *Nucleic Acids Res.*, **19**, 3153.
45. Penner, G. and Bezte, L. (1994). *Nucleic Acids Res.*, **22**, 1780.
46. Wiese, U., Wulfert, M., Prusiner, S., and Riesner, D. (1995). *Electrophoresis*, **16**, 1851.
47. Uitterlinden, A. G., Slagboom, P. E., Knook, D. L., and Vijg, J. (1989). *Proc. Natl. Acad. Sci. USA*, **86**, 2742.
48. Li, D. and Vijg, J. (1996). *Nucleic Acids Res.*, **24**, 538.
49. Vijg, J. (1995). *Mol. Biotechnol.*, **4**, 275.
50. Uitterlinden, A. and Vijg, J. (1994). *Two-dimensional DNA typing. A parallel approach to genome analysis*. Ellis Horwood, New York.
51. Van Orsouw, N., Li, D., VanderVlies, P., *et al.* (1996). *Hum. Mol. Gen.*, **5**, 755.
52. Steger, G. (1994). *Nucleic Acids Res.*, **22**, 2760.
53. Brossette, S. and Wartell, R. (1994). *Nucleic Acids Res.*, **22**, 4321.
54. Ophoff, R. A., Terwindt, G. M., Vergouwe, M. N., *et al.* (1996). *Cell*, **87**, 543.
55. Shiraishi, M., Lerman, L. S., and Sekiya, T. (1995). *Proc. Natl. Acad. Sci. USA*, **92**, 4229.

Cleavage using RNase to detect mutations

H. NAGASE and Y. NAKAMURA

1. Introduction

Recent advances in the human genome project will allow us to obtain information on the whole human genomic sequence and to map the positions of genes responsible for human hereditary diseases including both simple and complex genetic trait diseases. This information will provide a large number of candidate genes which will be used to identify the real responsible gene and assess its functions. For this purpose detection of mutations in genes is one of the most crucial techniques required. In order to screen for mutations, three methods have been frequently employed; detection of conformational changes between mutant and normal sequence, cleavage of mismatched heteroduplex, and direct sequencing. Conformational change detection, such as single-strand conformation polymorphism (SSCP) (1), denaturing gradient gel electrophoresis (DGGE) (2), and heteroduplex analysis (HA) (3), are most commonly utilized because of relatively easy handling and high sensitivity. However, there are some disadvantages in these methods. One main problem is a restriction of the fragment size for analysis; for SSCP this is less than 200 bp (4), for HA it is 200–300 bp, and for DGGE it is less than 600 bp (5). In SSCP and HA, sensitivity for special mismatches is condition-dependent, and DGGE require about 60 bp oligonucleotides with a G,C clamp to obtain high sensitivity (6). Chemical cleavage methods can analyse relatively long fragments, more than 1 kb. Among chemical cleavage methods, very early studies used S1 and Mung bean nucleases. However these nucleases frequently missed single base mismatches in heteroduplexes (7, 8). Another commonly used method is based on the principles of the Maxam and Gilbert sequencing technique (9). This method requires the biohazardous chemicals hydroxylamine and osmium tetroxide (HOT) to detect mutations (10). Although 100% sensitivity has been estimated, because of the use of hazardous chemicals and the multistep procedure, it has been performed in fewer laboratories.

The ribonuclease protection assay or RNase cleavage assay (RPA) has the same benefits of the other chemical cleavage method but without using hazardous chemicals. However, only 50–70% sensitivity for single base pair mismatch has been reported using the original method with RNase A (11). RPA has also been utilized to study other various aspects of RNA and DNA structure, such as detection and quantification of specific RNA transcripts (12), mapping transcribed sequences (13), measuring mRNA levels for promoter activity assay (14), and detection of RNase contamination (15). These methods have the advantage that the problem of template degradation or artefactual products in the procedure are small and that the results are highly reproducible. This advantage is also seen in the mismatch detection assay. Since recent modifications have improved the intricacy and sensitivity of this method, it may become more popular for the screening of mutation detection. In this chapter we discuss methodologies of RPA through our successful study of mutation detection of the *APC* gene in familial adenomatous polyposis (FAP) patients and somatic tumours (16), with useful modifications from other published literature.

2. RNase protection assay for mutation detection

RPA was used to detect RNA:DNA (11) and RNA:RNA mismatches (17) after the development of efficient *in vitro* RNA synthesis techniques (18). The method is theoretically very simple: RNA probes hybridized with specific DNA or RNA fragments are digested with RNase, which can recognize single-stranded RNA, and the digested probe is detected as shown in *Figure 1*. This method can analyse relatively large fragments and has good sensitivity without hazardous chemicals, if both sense and antisense are analysed or specifically optimized RNases are used. Recent techniques also allow us to map mutations in tested fragments within a day (19).

2.1 Evaluation of the sensitivity

The main problem with this methodology is that the sensitivity is not 100%. The original paper using RNase A for RNA:DNA heteroduplex reported that less than 50% of all single base mismatches could be identified (11). *Table 1* summarizes sensitivities of mismatch detection using the original RNase A for RNA:DNA heteroduplex and a combination of RNase A and RNase T1 for RNA:RNA heteroduplex (20); especially, single base pair mismatches on the purine RNA probes as they are not cleaved efficiently under normal conditions. However, according to the original paper a maximum of 88% of single base mismatches are detected using both strands of wild-type sequence as probes for DNA:RNA hetroduplexes. Although all mismatches have not been studied completely, RNA:RNA heteroduplexes may give better sensitivity. Even using both strands of RNA, some mismatches of UG(AC) and UC(AG) will not be detected in DNA:RNA heteroduplexes. In fact, in our study for the detection of *APC* mutations using RNase A, we have detected neither T

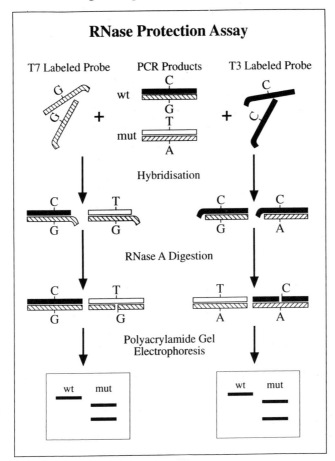

Figure 1. A summary of RNase protection assay.

to G transitions (U:C mismatch) nor the complimentary A to C transitions (A:G mismatch), while two of this type of *APC* germline mutations have so far been reported (*Table 2*). In *Figure 2* the strategy we used for *APC* mutation detection are summarized. Of course, in our case we do not know exactly whether no A to C or T to G transitions exist or if we have false negative results by RPA. However, no alterations were detected by SSCP analyses (75–90% sensitivity) of exons 6, 9, 15, or the 5′ part of exon 16 in any of the 28 FAP patients, whose mutations were not detected by RPA of the entire *APC* coding sequence and the flanking intron sequences. We then analysed the samples in the region between codon 999 and 1546 by direct sequencing and only one mutation (CGA–TGA) in codon 1450 was detected out of the 28 cases (21). *Table 3* summarizes the results of the mutation search using RPA. Over 60% of mutations which create truncated gene products were detected

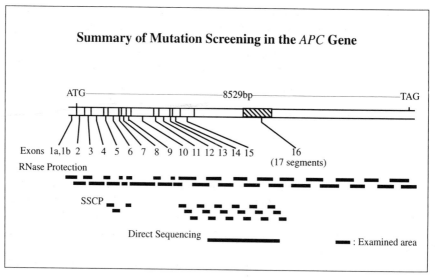

Figure 2. Strategy used to detect mutations in the *APC* gene.

Table 1. Cleavage of single base pair mismatches in RNA:DNA and RNA:RNA hetero-duplexes by ribonuclease A (adopted from refs 2 and 20)

RNA:DNA mismatch (complementary strand)			RNA:RNA mismatch		
Mismatch	**No. cleaved/No. tested**		**Mismatch**	**No. cleaved/No. tested**	
C:A (G:T)	22/22	(1/14)	G:U (——-)	13/156	(——-)
C:C (G:G)	4/4	(0/6)	C:C (G:G)	13/24	(9/9)
C:T (G:A)	9/9	(1/7)	C:U (——-)	37/40	(——-)
U:G (A:C)	7/11	(4/12)	—— (A:C)	——	(107/143)
U:C (A:G)	3/3[a]	(2/3)	—— (A:G)	——	(13/13)
U:T (A:A)	4/4	(2/6)	U:U (A:A)	11/11	(17/19)

[a] Only 5% of probes are cleaved.

Table 2. Single base pair mismatches due to *APC* gene mutations and polymorphisms detected by RPA

RNA:DNA mismatch	Germline mutation	Somatic mutation	Total (CpG site)	Polymorphisms	No. cleaved/No. tested (from Table 1)	
C:A or G:T	24	13	37 (20)	6	22/22	(1/14)
C:C or G:G	7	2	9	0	4/4	(0/6)
C:T or G:A	3	7	10	0	9/9	(1/7)
U:G or A:C	0	5	5	10	7/11	(4/12)
U:C or A:G	0	0	0	0	3/3[a]	(2/3)
U:T or A:A	5	2	7	2	4/4	(2/6)

[a] Only 5% of probes are cleaved.

Table 3. *APC* gene mutations detected in FAP and colorectal tumours by RPA

	Independent FAP families	Colorectal tumour Adenoma	Carcinoma	Total
Alteration detected	100	107	43	150
Total alleles examined	154	159[a]	76[a]	235
Percentage of detection	65%	67%	57%	64%

[a] 15 of 87 adenoma and 26 of 51 carcinoma have LOH in the *APC* locus.

in both FAP germline and colorectal tumours. We conclude that in the RPA, using RNase A 61 FAP germline mutations were detected between codons 999 and 1546, while 62 were detected in the same region when fully analysed by RPA, SSCP and direct sequencing. 11 out of 12 (92%) independent single base mismatch mutations and all 33 frame-shift mutations including five single base deletions or insertions were detected. For mutation searches in the *APC* gene, DGGE, SSCP and direct sequencing methods have been employed elsewhere. In studies using DGGE or direct sequencing, similar mutation detection rates were obtained in FAP germline and colorectal tumours, while SSCP did not reach the same rate (16).

Some recent modifications can improve the sensitivity of RPA. RNase I is an enzyme purified from *E. coli* and has a broad specificity capable of cleaving phosphodiester bonds involving all four nucleotides (22). Although sensitivity to detect the G:U pseudomismatch is still low (23), it can be used in place of RNase A and efficiency of RPA can be improved. RNase T2 may also be another option, which has non-specific endonuclease activity and optimal activity at low pH (24). A recent technique of non-isotopic RNase cleavage assay (NIRCA), using cleavage of RNA:RNA heteroduplexes immediately following *in vitro* RNA transcription from the PCR products, can also improve the sensitivity (19) (see later). In blind studies for the NIRCA, it was found that 90% of single base mismatches were detected in homozygous and 88% were detected in heterozygous targets (analysing a 585 bp fragment) (19).

2.2 Source material

Genomic DNA and messenger RNA are generally used for this type of study. Although both are able to be utilized as the heteroduplex template directly hybridized with the RNA probe, PCR amplified fragments from the templates are commonly used. Since RNase protection assay was able to detect as few as 10^5–10^7 molecules, 20–50 ng of cloned plasmid DNA, 3–6 μg of genomic DNA, or 1–10 μg of total mRNAs have been utilized for RPA. Another point to remember is that mutations do not occur only in the coding sequence, but also they can occur in the non-coding region. In our case, three splice site mutations and a 1 bp deletion in the intron have been detected. These mutations can not be detected by searches of mRNA. On the other hand, mutations within PCR primers will not be detected, since the mutated sequence is not amplified from the experimental samples and also mismatches close to the end of the amplified sequence may be difficult to detect. We have designed amplimers for PCR in intron sequences flanking a minimum of about 40 bp from the exon–intron junction and 17 segments in the last large exon with at least 50 bp overlap in each set of primers (*Figure 2*). Although PCR products less than 500 bp are used in our *APC* study, 1 kb of PCR fragments can be analysed in this way. PCR artefacts are a major problem for conformational change detection strategy, while in RPA common artefact bands from other genomic sequence have little problem throughout the standard RPA experiments, if the proper sequence is amplified. It will, however, become crucial for RNA synthesis using PCR products with RNA polymerase promoter.

2.3 PCR for RNase protection assay

The PCR for RPA does not require strictly optimized conditions as for SSCP, DGGE, and HA. Amplified products from other genomic sequences or cDNAs are usually not hybridized with the RNA probe and do not protect the probe from RNase digestion. In PCR reactions the normal sequence may be accidentally amplified from contaminating clones or PCR products. This is a major problem and must be avoided. A negative control must be used in all PCR procedures and electrophoresed in agarose gels with several PCR products to determine the product size and the degree of contamination. A PCR reaction without any DNA sample must be performed and used for the RPA experiment as a negative control. Misincorporation by *Taq* polymerase is another potential problem. However, we use normal *Taq* and repeat the RPA using other PCR products amplified by independent PCR reactions.

Protocol 1. A standard PCR amplification of sample DNAs[a]

Equipment and reagents

- Genomic DNA
- 10 × PCR buffer: 166 mM NH_4SO_4, 670 mM Tris pH 8.8, 20 mM $MgCl_2$, 100 mM β-mercaptoethanol, and 67 μM EDTA in HPLC grade H_2O

- 25 mM dNTP
- *Taq* polymerase
- Thermal cycler

Method

1. Genomic DNA samples from patients and tumours prepared by a proper method are diluted to 50 ng/μl in TE. 2 μl of 50 ng/μl genomic DNA are used for 25 μl PCR reaction. 2 μl of 50 fg/μl plasmid DNA including wild-type sequence and 2 μl of dH_2O are also used as a positive control and a negative control respectively.

2. Prepare reaction mixture of 2.5 μl 10 × PCR buffer, 2.5 μl DMSO, 1.5 μl 25 mM dNTPs, 1.5 μl 78 mM MgCl$_2$, 0.25 μl of each 100 pmol/μl PCR primer, 14.25 μl dH$_2$O, and 0.25 μl 5 U/μl *Taq* polymerase per sample.

3. Aliquot 23 μl reaction mixture into 0.2 ml PCR tubes.

4. Add 2 μl 50 ng/μl DNA to each tube, mix with positive displacement tip or filtered tip.

5. Add one drop of mineral oil to each tube, if necessary.

6. Typical thermal cycle profile:[b]

   (a) Hold 2 min at 94°C.

   (b) 35 cycles of 94°C 30 sec; 55°C 30 sec; 72°C 40 sec.

   (c) Hold 4 min at 72°C.

   (d) Hold indefinitely at 4°C.

7. 4 μl of the PCR products are electrophoresed in a 2% agarose gel with ethidium bromide to confirm the results. Each 1 μl product is used for following RPAs. If abnormal bands are found in the RPA, the remainder of the products are used for subsequent sequencing to detect mutations.

[a]This protocol is only an example. Not every combination of template sequence and primer set has always worked under these conditions. The type of thermal cycler and *Taq* polymerase used may also influence the results.
[b]This is an example of amplification 400–500 bp products by primers (T_m = 58°C) for Perkin Elmer model 9600. PCR conditions should be modified for different PCR and the other thermal cyclers used.

Genomic DNA or messenger RNA can also be used directly as a template hybridized with an RNA probe. PCR products are recommended, because in some cases the mishybridizations between the RNA probe and pseudogenes or gene families can be avoided. Therefore the PCR fragments amplified from the target gene increase the specificity of the RPA.

2.4 RNA probe preparation

Efficient *in vitro* synthesis of the RNA hybridization probe has been successful using plasmids with an SP6 promoter (18). Based on this method the RNA probe transcribed from wild-type sequence has been used for RPA since first described in 1985. This standard method requires a subcloning step of the wild-type sequence into a plasmid. PCR-based *in vitro* transcription without subcloning into plasmids has been reported by H. Yang and P. W. Melera (25). A SP6 or T7 RNA polymerase promoter sequence was added to the 5' end of the antisense primer. PCR amplification was performed using a sense

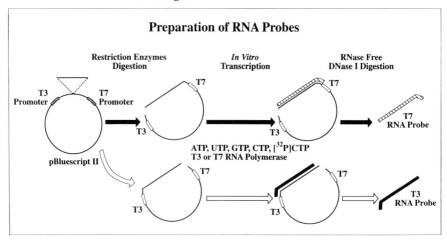

Figure 3. A summary of RNA probe preparation.

primer and the antisense primer with a promoter sequence, and an RNA probe was directly transcribed *in vitro*. A set of both PCR primers with either of the promoter sequence at 5′ end have been used for the second PCR reaction based on the nested PCR approach. The products can be utilized for either RNA probe synthesis for standard RPA, or RNA duplex synthesis for direct detection by agarose gel with ethidium staining after RNase digestion (19). The latter modification was called non-isotopic RNase cleavage assay (NIRCA).

2.4.1 Standard RNA probe preparation

For large scale studies, the standard method using a subcloned plasmid has some advantages despite the requirement for the multiple step procedure. In repeat experiments, constant levels of RNA synthesis are obtained from either wild-type sense or antisense sequence for probe preparation and for RNA duplex study using NIRCA, without having to make large amounts of PCR products amplified from a wild-type sequence in each case. Amplified PCR products from a wild-type sequence are subcloned into plasmids which have multiple cloning sites between different promoter sequences for either T3, T7, or SP6 RNA polymerase promoters (see *Figure 3*). Although the SP6 RNA polymerase is slightly more stable than T3 and T7 RNA polymerases, no major advantage is seen using either T3, T7, or SP6 promoters. For the *APC* study, we used pBluescript II SK(−) which has T3 and T7 promoters for standard blunt-end cloning into the *Eco*RV cloning site. Several rapid PCR fragment cloning systems are also now commercially available.

Protocol 2. RNA probe preparation for both strands by the subcloned plasmid

In order to obtain better sensitivity, RNA probes corresponding to both strands of wild-type sequence should be synthesized separately. *Figure 3* is a summary of the labelling procedure for subcloned plasmids.

Equipment and reagents

- Standard molecular biology reagents
- 5 × transcription buffer: 200 mM Tris–HCl pH 7.9, 30 mM MgCl$_2$, 10 mM spermidine, 50 mM NaCl
- dNTPs
- [α-^{32}P]CTP

- SP6, T3, or T7 RNA polymerase
- Phenol:chloroform:isoamyl alcohol (25:24:1)
- Hybridization buffer: 45 mM Pipes pH 6.4, 1.125 mM EDTA, 0.45 M NaCl, 90% deionized formamide
- Centrifuge

A. *Template plasmid DNA preparation*

1. The PCR products amplified from wild-type sequence are subcloned into a plasmid with RNA polymerase promoters.

2. A single colony is picked under the colour selection after transformation into *E. coli* and the plasmid DNA is prepared.

3. Linearized plasmid DNA is prepared by digestion with a restriction enzyme recognizing a site between one promoter and the insert for the sense probe, and another plasmid DNA is linearized by those between another promoter and the insert (see *Figure 3*). The restriction enzyme should leave a 5' overhang.[a]

4. After determination of complete digestion, each DNA is prepared to 250 ng–1 μg/μl (1 μg for more than 100 samples).

B. *Labelling of the RNA probe*

1. Add the following reagents in the order listed:

   - 5 × transcription buffer   4 μl
   - 100 mM DTT   2 μl
   - 3.3 mM ATP UTP GTP mix   3 μl
   - 1 mM CTP   1 μl
   - Template DNA (250 ng/μl) in TE[b]   4 μl
   - [α-^{32}P]CTP (800 Ci/mmol)   4 μl
   - Ribonuclease inhibitor (20–40 U/μl)   1 μl
   - SP6, T3, or T7 RNA polymerase (10–20 U/μl)   1 μl
   - Final volume   20 μl

2. Incubate for 45 min at 37°C.

3. Add 2 μl RNase-free DNase I (1 U/μl) and incubate at 37°C for 15 min to digest template DNA completely.[c]

4. Add 20 μl 0.2 M EDTA to terminate the reaction and 160 μl of TE.

Protocol 2. *Continued*

5. Extract with the same volume of TE-saturated phenol:chloroform: isoamyl alcohol (25:24:1). Vortex for 1 min and centrifuge at 10 000 *g* for 5 min.

6. Put the supernatant into a fresh tube and add 2 μl glycogen (10 mg/ml), 50 μl 8 M NH₄OAC, and 500 μl 100% ethanol. Mix well and centrifuge at 14 000 *g* for 10 min at 4°C.

7. Remove the supernatant and wash the pellet with 500 μl of 70% ethanol. Dry the pellet under vacuum.

8. Resuspend in 50 μl of hybridization buffer.

9. Take 1 μl to determine the amount of incorporated radioactivity[d] and store the rest at −70°C.

[a] If there is no choice and the plasmid DNAs must be linearized by a enzyme to yield a 3′ overhang, the overhang sequence must be filled in by Klenow DNA polymerase to create blunt-ends to reduce the yield of complementary strands. Otherwise sense and antisense RNA are transcribed from both strands and reduce the sensitivity of the assay.
[b] Total 1 μg of the linearized plasmid DNAs or 0.2–1 μg of PCR products (see *Protocol 3*) are used as a template.
[c] Incomplete digestion of complementary DNA sequence will protect RNA probes from RNase cleavage at a later stage. This gives the same pattern of constant bands by autoradiography in all samples, including negative controls.
[d] 2–10 × 10⁵ c.p.m./μl of incorporation is normally expected in this protocol (for 100–600 RPA reaction).

2.4.2 PCR-based RNA probe preparation

In the first descriptions of this procedure, a single PCR step was performed using a sense primer and an antisense primer with either an SP6 or a T7 RNA polymerase promoter sequence. Nested PCR using two pairs of outside primers and inside primers with the promoter sequences is now frequently preferred to amplify the template sequence.

Protocol 3. RNA probe preparation for both strands from PCR fragments

Both first and second PCR conditions are described in *Protocol 1*.

Method

1. Genomic DNAs (50 ng–1 μg per sample) are amplified by the outer primer set with 25–30 amplification cycles following *Protocol 1*. This step may be avoided in some cases, where the products are amplified properly by only a set of the inner primers.

2. 1–10 μl of the first PCR products or genomic DNAs (50 ng–1 μg per sample) are amplified by the inner primer set.

3. 1–4 µl of the PCR products will be utilized for 20 µl transcription reaction.

4. RNA probes are prepared by labelling procedures in *Protocol 2*. These PCR products are also transcribed for RNA duplex study of NIRCA without radioisotopes.

In order to avoid cumbersome steps of subcloning, PCR products with either SP6, T7, or T3 RNA polymerase promoter sequences can be used directly for *in vitro* RNA probe preparation. The promoter sequence is attached to the 5' end of the sense and antisense primer. The primer sequence consists of about 20 bases of T7 or SP6 promoter consensus sequence at the 5' ends, followed by 15–18 bases of target-specific sequence (*Figure 4*). For the nested PCR another pair of outer primers corresponding the target sequence are designed with a suitable software program.

2.5 RNase protection

Several RNases, such as RNase A, RNase T1, RNase T2, and RNase I, have been used for RPA. It is possible to use a combination of these enzymes for

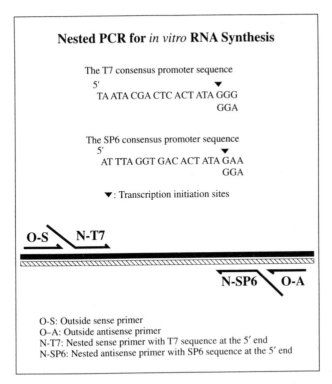

Figure 4. PCR primers designed for PCR-based *in vitro* RNA synthesis.

RPA, however, RNase T2 acts at a lower and RNase I at a higher optimum pH than RNase A and T1. Since RNase T1 and T2 do not cleave mismatches which can not be cleaved by RNase A in normal conditions, addition of neither RNase T1 nor RNase T2 can improve mismatch detection in RNase A cleavage of DNA:RNA heteroduplexes. If the mismatched hybrids are digested under three separate sets of conditions by each of the three enzymes, better sensitivity will be obtained, but even this additional work only increases detection rate by a relatively small amount.

In the RNase protection assay RNA probes corresponding to wild-type sequence are hybridized with the target sequence amplified from samples, and the hybrids are digested by RNase. RNase cleaves the RNA probe at the mismatched sites and protected RNA fragments are detected by autoradiography following electrophoresis. *Figure 1* shows a summary of the assay.

Test reactions are performed prior to every large scale study. In the test reaction, several PCR products from samples and wild-type, and two negative controls are used for the RPA reaction. The reaction and the RNA probe used are checked by electrophoresis and autoradiography.

Protocol 4. RNase protection assay

Equipment and reagents

- Hybridization buffer: 45 mM Pipes pH 6.4, 1.125 mM EDTA, 0.45 M NaCl, 90% deionized formamide (store at –20°C)
- RNase buffer: 0.2 M NaCl, 0.1 M LiCl, 0.02 M Tris–HCl pH 7.5, 1 mM EDTA (store at –20°C)
- 1 mg/ml RNase A: boil for10 min to remove DNase activity (store at 4°C)
- Proteinase K/10% SDS: 5 mg/ml proteinase K in 10% SDS (store at –20°C)
- Phenol:CIAA: TE-saturated phenol:chloroform:isoamyl alcohol (25:24:1) (store at 4°C)

- Water-baths at 94°C and 37°C
- Stop buffer: 0.3% (w/v) xylene cyanol, 0.3% (w/v) bromophenol blue in formamide (store at 4°C)

Samples

- 1 μl PCR products amplified from samples
- 1 μl PCR products amplified from dH$_2$O instead of a sample as a negative control
- 1 μl dH$_2$O as another negative control
- 1 μl PCR products amplified from wild-type sequence

A. *Hybridization*

1. Put 1 μl of the PCR products carefully into the bottoms of 1.5 ml microcentrifuge tubes.

2. Add 8 μl of hybridization mixture avoiding touching the bottom of the tubes: RNA probe 150 000 c.p.m. in 8 μl hybridization buffer. (If 7.5 × 10^5 c.p.m./μl of the radioactivity have been incorporated in the RNA probe, for 100 samples, 20 μl of RNA probe are diluted in 800 μl of hybridization buffer.)

3. Vortex and spin down. Denature PCR products at 94°C for 4 min.

4. Hybridize at 50°C for 2 h, or overnight.

B. *RNase cleavage*

1. Add 90 μl RNase: 1 mg/μl of RNase A is 40 times diluted by RNase buffer. (250 μl of 1 mg/μl of RNase A are added to 10 ml RNase buffer for 100 reactions.)

2. Vortex and spin down. Cleave mismatched RNA probe at 37 °C for 45 min.

3. Add 11 μl of proteinase K/10% SDS.

4. Vortex and spin down. Digest RNase A at 37 °C for 45 min.

5. Add 110 μl phenol:CIAA. Vortex for 5 min and centrifuge at 10 000 *g* for 2–3 min.

6. Put 20 μl of supernatant into microcentrifuge tubes which contain 20 μl of stop buffer in advance.

7. Vortex and spin down. Samples are ready for electrophoresis or stored at –70 °C.

We do not normally precipitate the extracted products. It does not affect the electrophoresis and sensitivity of detection.

2.6 Detection of digested probe

Protected RNA probe can detected by autoradiography following denaturing polyacrylamide gel electrophoresis. The map position of the mismatch is estimated from the size of protected products.

8% polyacrylamide 8 M urea gel cast into a 15 cm width, 13 cm height, and 1 mm thick plate with 20 teeth comb for vertical slab gel apparatus is used for approximately 500 bp RNA probe. This gel can be manipulated more easily than a sequence gel. Since RNA has slightly slower electrophoretic mobility than DNA under these conditions, the actual size of the RNA fragments were found to be about 5% smaller than the size indicated by the DNA fragment.

Protocol 5. Denaturing polyacrylamide gel electrophoresis

Equipment and reagents

- Vertical slab gel apparatus
- 8 M urea/8% polyacrylamide
- TBE
- 10% methanol/10% acetic acid
- Whatman 3MM paper
- Gel dryer
- X-ray film

A. *Electrophoresis*

1. Denature samples obtained in *Protocol 4* at 94 °C for 4 min.

2. Load 15 μl of denatured samples on an 8 M urea/8% polyacrylamide gel and perform electrophoresis under constant 250 V using 1 × TBE buffer for 1.5–2 h.

Protocol 2. *Continued*

3. Stop the electrophoresis 5 min after the bromophenol blue reaches the bottom of the gel.[a]

4. Transfer the gel to a bath containing 10% methanol and 10% acetic acid in water. Fix for 1 h.

5. Pick up the gel and cover with Whatman 3MM paper. Lay a plastic wrap on top of the gel.

6. Dry the gel for 1–2 h under vacuum on a commercial gel dryer at 80°C.

B. *Autoradiography*

1. Put gels into an autoradiograph cassette with X-ray film.[b]

2. Expose the film at –70°C with an intensifying screen overnight.[c]

3. Develop the film.

[a]Everything larger than 20 bp of cleaved RNA probe should remain in the gel. In the case of 8 M urea/6% polyacrylamide gel electrophoresis should be stopped immediately after BPB runs off.
[b]Normally eight gels (160 samples) are in a large cassette.
[c]Another short exposure may be needed for large cleaved RNA probe fragments.

Abnormal bands corresponding to cleaved RNA probes at mismatched positions are seen in autoradiograms. An abnormal pattern of a lane is easily recognized (see *Figure 5A*). For germline mutations, the abnormal patterns result from mutations in half of the PCR products, because of the presence of both normal and mutant alleles. Furthermore some single base mismatches are not cleaved completely by RNases and may not be visualized as abnormal bands in a short exposure. Occasional faint abnormal bands should be checked in the next step of mutation detection. However, most of these faint bands correspond to PCR errors due to *Taq* polymerase misincorporation. Therefore, these abnormal bands will frequently disappear in another RPA using independent PCR products.

2.7 Mutation detection by sequencing of the PCR products

PCR products which have given rise to an abnormal pattern on the autoradiogram are then analysed by sequencing for the actual mutation. Any sequencing procedure can be employed to determine the mutant sequence in the PCR fragment used for the RPA. Determination of the whole sequence in the fragment is very time-consuming, especially when there is a large number of samples. In order to reduce the labour of this time-consuming step, the following points must be checked before the sequencing reactions are performed. The results obtained by RPA should suggest the approximate map positions of mutations. According to this map, the sequences are checked carefully around these positions. Secondly, sequencing samples using the same sequencing primer are electrophoresed at the same time as in *Figure 5B*. The picture is an example of loading lane 1, 3, 5, and 7, with G, A, T, and C

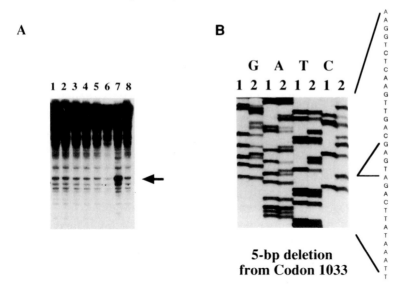

Figure 5. (A) An example of an autoradiogram showing *APC* germline mutation by RNase protection assay and (B) sequencing of the same samples.

respectively from the same sample, and, 2, 4, 6, and 8 are from another sample in the same order. In this way the mutation pattern is easily detected at the exact position without having to read whole sequences.

Protocol 6.   Mutation detection by sequencing

1. The rest of PCR products which are used for the RPA are subcloned into a plasmid. After transformation more than 50 white colonies are picked by colour selection and DNAs are prepared from the mixed population. (This step can be ignored for direct sequencing from PCR products.)

2. The subcloned plasmid DNAs or PCR products are sequenced using a dideoxy sequencing method.

3. Samples using the same synthetic primer are loaded simultaneously; dideoxy GTP terminator tubes from all the samples which occupy the first lanes and then tubes including the same dideoxy terminators are loaded in the order ATC.[a]

4. Mutations are detected easily as abnormal patterns after autoradiography. Sequences need only be read at these specific regions.

[a] In our case, a maximum of five samples were loaded simultaneously, because of the difficulty of reading from the far distance of GATC. Frame-shift mutations are relatively difficult to read and in cases where all of the mutations are frame-shifted, another single sample loading might be needed for reading a mutant sequence.

2.8 Other modified methodologies for mutation detection

As mentioned before, several useful modifications have been published and are commercially available. Some of them have already been included in the above protocols. Here we touch on the use of RNase I and NIRCA. RNase I cleavage may give some improvement of the RPA. Although some G:U pseudomismatches can not be cleaved by this enzyme, it may increase the detection rate of the RPA compared with RNase A. One benefit is that RNase I is inactivated more easily than RNase A. It does not require phenol extraction following proteinase digestion to remove RNases and is inactivated by 0.1% sodium dodecyl sulfate (SDS) (23).

NIRCA, which was recently developed by Marianna Goldrick (19), has several advantages; no radioisotope requirement, agarose gel detection protocol, high sensitivity (90% for single base mismatches in homozygous and 88% in heterozygous targets), and a quick method (results obtained within a day). *Figure 6* summarizes the NIRCA procedure.

Protocol 7.   Non-isotopic RNase cleavage assay (NIRCA)[a]

1. PCR amplification of experimental and wild-type control sequences (follow *Protocol 3*, omitting the radioisotope).

2. Transcription of both strands of experimental and wild-type PCR products (follow *Protocol 3*, omitting the radioisotope).

3. Hybridization of complementary experimental and wild-type transcripts. Denature the mix of wild-type and the complementary experimental RNA transcripts at 90–95°C for 3 min and cool to room temperature.[b]

4. Digestion of RNA duplexes with RNase. Add 16 μl of diluted RNase solution containing ethidium bromide to each 4 μl aliquot of target RNA.[c] Incubate at 37°C for 45 min.

5. Separation of digestion products by gel electrophoresis in agarose. Load the sample with 4 μl of loading buffer on a 2% agarose gel in 1 × TBE.[d] Examine the gel on a UV transilluminator.

[a] More details can be obtained from Ambion Inc.
[b] No hybridization time is necessary and the hybridized product can be stored at −20°C.
[c] Two different solutions should be used in independent reactions to obtain high sensitivity.
[d] Higher percentage agarose gels may be useful for analysis of short target regions (less than 400 bp).

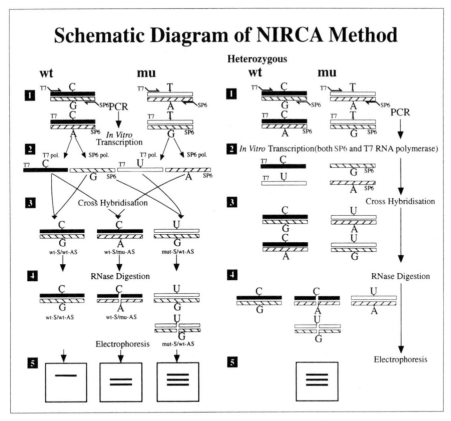

Figure 6. A summary of non-isotopic RNase cleavage assay (NIRCA).

Acknowledgements

We thank Professor Allan Balmain and Dr Marianna Goldrick for useful comments. We also wish to acknowledge Dr Isamu Nishisho, Dr Yasuo Miyoshi, Dr Hiroshi Ando, and Dr Takahiro Mori for many contributions in this work. H. N. is supported by the Cancer Research Campaign.

References

1. Orita, M., Iwahana, H., Kanazawa, H., Hayashi, K., and Sekiya, T. (1989). *Proc. Natl. Acad. Sci. USA*, **86**, 2766.
2. Myers, R. M., Lumelsky, N., Lerman, L. S., and Maniatis, T. (1985). *Nature*, **313**, 495.
3. Nagamine, C. M., Chan, K., and Lau, Y. F. C. (1989). *Am. J. Hum. Genet.*, **2**, 425.
4. Sarkar, G., Yoon, H-S., and Sommer, S. S. (1992). *Nucleic Acids Res.*, **20**, 871.
5. Grompe, M. (1993). *Nature Genet.*, **5**, 111.

6. Sheffield, V. C., Cox, D. R., Lerman, L. S., and Myers, R. M. (1989). *Mol. Cell Biol.*, **7**, 4024.

7. Shenk, T. E., Rhodes, C., Rigby, P. W., and Berg, P. (1975). *Proc. Natl. Acad. Sci. USA*, **72**, 989.

8. Kowalski, D., Kroeker, W. D., and Laskowski, M. (1976). *Biochemistry*, **15**, 4457.

9. Maxam, A. M. and Gilbert, W. (1977). *Proc. Natl. Acad. Sci. USA*, **74**, 560.

10. Coton, R. G. H, Rodrigues, N. R., and Campbell, R. D. (1988). *Proc. Natl. Acad. Sci. USA*, **85**, 4397.

11. Myers, R. M., Larin, Z., and Maniatis, T. (1985). *Science*, **230**, 1242.

12. Trudel, M., Magram, J., Brudckner, L., and Costantini, F. (1987). *Proc. Natl. Acad. Sci. USA*, **82**, 7575.

13. Shimada, T., Inokuchi, K., and Nienhuis, A. W. (1987). *Mol. Cell Biol.*, **7**, 2830.

14. Sylvers, L. A., Maedor, J., and Winkler, M. M. (1994). *FASEB J.*, **8**, A1276.

15. Bolger, R. and Thompson, D. (1994). *Am. Biotechnol. Lab.*, **12**, 113.

16. Nagase, H. and Nakamura, Y. (1993). *Hum. Mutat.*, **2**, 425.

17. Winter, E., Yamamoto, F., Almoguera, C., and Perucho, M. (1985). *Proc. Natl. Acad. Sci. USA*, **82**, 7575.

18. Melton, D. A., Krieg, P. A., Rebagliati, M. R., Maniatis, T., Zinn, K., and Green, M. R. (1984). *Nucleic Acids Res.*, **12**, 7035.

19. Goldrick, M. M., Kimball, G. R., Liu, Q., Martin, L. A., Sommer, S. S., and Tseng, J. Y-H. (1996). *BioTechniques*, **21**, 106.

20. Aranda, M. A., Fraile, A., Farcia-Arenal, F., and Malpica, J. M. (1995). *Arch. Virol.*, **140**, 1373.

21. Mori, T., Nagase, H., Horii, A., Miyoshi, Y., Shimano, T., Nakatsuru, S., *et al.* (1994). *Genes Chromosomes Cancer*, **9**, 168.

22. Meador III, J., Cannon, B., Cannistraro, V. J., and Kennell, D. (1990). *Eur. J. Biochem.*, **187**, 549.

23. Murthy, K. K., Shen, S-H., and Banville, D. (1995). *DNA Cell Biol.*, **14**, 87.

24. Saccomonno, C. F., Bordonaro, M., Chen, J. S., and Nordstorm, J. L. (1992). *BioTechniques*, **113**, 847.

25. Yang, H. and Melera, P. W. (1992). *BioTechniques*, **13**, 922.

5

Cleavage of mismatched bases using chemical reagents

FRANCESCO GIANNELLI, PETER M. GREEN, and
SUSAN J. RAMUS

1. Introduction

Chemical cleavage of mismatch (CCM) was developed in 1988 as a method for screening cloned pieces of DNA for single base mismatches (1), but in combination with PCR was very soon used for the direct analysis of genomic DNA (2) and, later, for the analysis of mRNA (3). This technique detects mismatches in hybrid DNA. Heteroduplexes are formed by adding labelled normal DNA usually to an excess of unlabelled patient DNA. The hetero-duplexes are reacted with either hydroxylamine which recognizes mismatched C bases or osmium tetroxide which recognizes mismatched T bases. Any mis-matched or unmatched C or T bases are modified by the respective chemicals so that the modified strand can be cleaved with piperidine at the site of the mismatch. The products are analysed by electrophoresis on a denaturing polyacrylamide gel and the size of the fast running cleavage bands indicates the site of the mismatch. All manipulations involving the chemicals must be performed in a fume-hood.

The method shares, with denaturing gradient gel electrophoresis and related methods, the ability to detect virtually any sequence change but in addition it does not require special expensive primers or enzymes, is not very sensitive to sequence contexts, it can screen larger fragments (1.5–1.8 kb) and, most importantly, indicates the location of mutations within the segments analysed. Although more complex than simple procedures such as SSCP, it is not only more effective but more predictable than this and related techniques and results in much faster screening procedures whenever a gene with many exons can be examined at the RNA level. Not surprisingly therefore this method has formed the basis of the first procedures for the detection of all mutations in large and complex genes (4, 5). The CCM method has been used in many laboratories. A search of the Science Citation Index database showed that between 1988 and 1996 there have been at least 95 papers published describing the detection of mutations with this method. There are many possible variations of the CCM method that may suit different situations and it can be applied to DNA and DNA:RNA heteroduplexes.

The experiments can be performed with different types of heteroduplexes:

(a) Only the normal control labelled (2).

(b) Both patient and normal control labelled (6).

(c) Only the patient labelled (7) (this is useful if the patient is known to be heterozygous for the unknown mutation, i.e. dominantly inherited diseases).

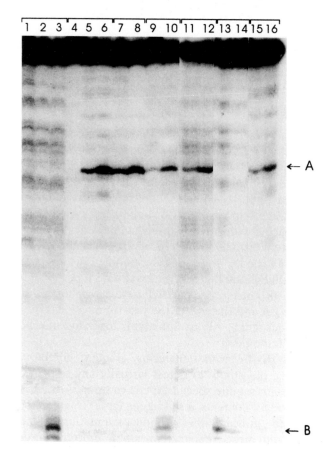

Figure 1. Traditional CCM and single tube CCM using a ^{32}P end-labelled probe. The hydroxylamine and osmium tetroxide cleavage products are indicated by an A and B respectively. Lanes 1 to 6 are the traditional CCM with 0, 1, and 5 min OsO_4 and 0, 20, and 60 min hydroxylamine respectively. Lanes 9 and 10 are the single tube CCM (hydroxylamine then OsO_4). Cleavage was detectable using both OsO_4 and hydroxylamine in lane 10 with 60 min hydroxylamine then 5 min OsO_4. Lanes 7 and 8 are OsO_4 then hydroxylamine and lanes 11 and 12 are the two chemicals together. In both of these sets of reactions only the hydroxylamine reaction was detectable. Lanes 13 to 16 are controls for dilutions of the chemicals. Reproduced from ref. 15 with permission.

The DNA strands can be labelled in many different ways:

(a) ^{32}P end-labelling (2) (labelling primers or PCR products).
(b) ^{32}P uniform labelling (8).
(c) ^{35}S uniform labelling (9).
(d) Unlabelled (with silver staining of the polyacrylamide gel) (10).
(e) Biotin labelling (11).
(f) Fluorescent uniform labelling (12).
(g) Fluorescent end-labelling (12–14).

Time-saving procedures have been developed:

(a) Performing both chemical reactions in the same tube (15).
(b) Using solid phase techniques instead of precipitations (12, 16, 17).
(c) Multiplexing using an ABI automated sequencer (12, 13).

Thus the speed of the procedure has been increased by at least tenfold. Here we describe the basic procedure using ^{32}P-labelling with separate and successive chemical treatments followed by the solid phase fluorescent procedures based either on uniform or end-labelling. Examples of both ^{32}P end-labelled traditional CCM and single tube CM are given in *Figure 1*.

2. Basic procedures

Protocol 1. Chemical cleavage of mismatch (CCM) using a ^{32}P end-labelled probe

Equipment and reagents

- Fume-hood which can be used for radioactive chemicals
- β-counter
- Microcentrifuge
- [γ-^{32}P]dCTP
- Hydroxylammonium chloride (Fluka, Aldrich, or BDH AnalaR)
- Osmium tetroxide (Aldrich)
- Piperidine (Fluka)
- Pyridine (HPLC grade from Aldrich)
- 2 × annealing buffer: 6 ml 2 M NaCl, 120 μl 1 M Tris–HCl pH 7.5, 140 μl 1 M MgCl$_2$, 2.84 ml dH$_2$O
- Hydroxylamine solution: add 1.39 g of solid hydroxylammonium chloride to a glass tube and dissolve in 1.6 ml of dH$_2$O; add 1 ml of diethylamine slowly and then a further 750 μl. Test the pH of the solution by placing two drops on a piece of pH paper; the pH should be between 6–7. Store at 4°C.

- Diethylamine (Fluka or Aldrich)
- 4% osmium tetroxide solution (available from Aldrich)
- 10 × osmium tetroxide buffer: 100 mM Tris–HCl pH 7.7, 10 mM EDTA, 15% pyridine
- HOT stop buffer: 0.3 M NaAc pH 5.2, 0.1 mM EDTA, 25 μg/ml tRNA
- tRNA: 50 mg/ml stock solution (Baker's yeast, Boehringer Mannheim) as carrier. (Clean up the tRNA by phenol and chloroform extraction and precipitate with ethanol. Dissolve in dH$_2$O and heat treat at 95°C for 10 min. Store at –20°C.) Alternatively, use glycogen (Boehringer Mannheim) at 4 mg/ml final concentration.
- Loading dye: 95% formamide, 20 mM EDTA, 0.05% bromophenol blue, 0.05% xylene cyanol
- End-labelled DNA molecular weight marker (usually ϕX174 digested with HaeIII)

83

Protocol 1. *Continued*

A. *^{32}P end-labelled probe preparation*

1. Amplify by PCR the fragment of interest from DNA samples of a normal control and of the patients to be tested for mutations.

2. Gel purify the amplified PCR product from normal control DNA,[a] to remove any non-specific product and excess primers.

3. Precipitate the purified PCR product and resuspend in 10 μl of dH$_2$O.

4. Examine the quality and quantity of the purified PCR product by electrophoresis of 1 μl of the 10 μl of purified PCR product on a 1% agarose gel next to a molecular weight size marker (for example φX174 digested with *Hae*III).

5. Remove approx. 100 ng of DNA to a 1.5 ml microcentrifuge tube. Add 5 μl of 10 × polynucleotide kinase (PNK) buffer, and dH$_2$O to a total volume of 48 μl. Keep the tube on ice. Working in a radioactive area, add 1 μl (10 U) of polynucleotide kinase (PNK) and 1 μl of [γ-^{32}P]dCTP (3000 Ci/mmol). Incubate the tube at 37°C for 60 min.

6. Precipitate the probe. Add 5 μl of 3 M NaAc pH 5.2 and 125 μl of ice-cold 100% ethanol, and leave the tube at –20°C for 30 min. Centrifuge the tube for 15 min at 13 000 r.p.m. Remove the supernatant and re-suspend the pellet in 100 μl of dH$_2$O. Precipitate the probe again by adding 10 μl of 3 M NaAc pH 5.2 and 250 μl of ice-cold 100% ethanol, and leave the tube at –20°C for 30 min. Centrifuge the tube for 15 min at 13 000 r.p.m. Remove the supernatant and wash the pellet with 70% ethanol.

7. Count the dry pellet[b] in a β-counter and resuspend the pellet at 1000 c.p.m./μl in dH$_2$O.[c]

8. Check that the probe is properly resuspended.[d] Remove all of the dH$_2$O with the Gilson and hold it in front of a hand-held monitor. Compare the amount of radioactivity with that still on the side of the tube. Ensure that at least half of the counts are in the dH$_2$O. Place the dH$_2$O back into the same tube.

B. *Heteroduplex formation*

1. Add approx. 500 ng of unlabelled PCR product from each of the patients and the normal control[e] to separate 1.5 ml microcentrifuge tubes. To each tube add 50 μl (50 000 c.p.m.) of probe. Mix the DNA and add dH$_2$O to a total volume of 100 μl. Add 100 μl of 2 × annealing buffer to the tubes and mix.

2. Place a separate lid (with a small hole made with a needle) on the microcentrifuge tube.[f] Boil the tubes for 5 min, place on ice and replace the original lid, and then immediately incubate the tubes at 42°C for 1 h.

3. Precipitate the DNA by adding 20 μl of 3 M NaAc pH 5.2 and 500 μl of ice-cold 100% ethanol, and leave the tubes at –20°C for 30 min. Centrifuge the tubes for 15 min at 13 000 r.p.m. Remove the supernatant and wash the pellets with 70% ethanol.

4. Count the dry pellets[b] in a β-counter, and resuspend the pellets at 1000 c.p.m./μl in dH$_2$O.[g]

C. *Chemical modification reactions*

1. For each heteroduplex aliquot 6 μl (6000 c.p.m.) into six separate 1.5 ml microcentrifuge tubes. Label each tube with the heteroduplex number and either 0, 1, 5 min osmium tetroxide, 0, 20, 60 min hydroxylamine.

2. Perform all of the following manipulations in a fume-hood.

3. To each of the 1 h hydroxylamine tubes add 20 μl of the hydroxylamine solution and place in a 37°C heating block for 1 h. To each of the 20 min hydroxylamine tubes add 20 μl of the hydroxylamine solution and place in a 37°C heating block for 20 min.

4. Stop the reactions by adding 200 μl of HOT stop buffer and 750 μl of ice-cold 100% ethanol. Invert the tubes and place immediately at –20°C for at least 30 min.

5. To each of the 0 min hydroxylamine tubes add 200 μl of HOT stop buffer and 750 μl of ice-cold 100% ethanol. Add 20 μl of the hydroxylamine solution and invert the tubes, and place immediately at –20°C for at least 30 min.

6. Freshly dilute the 4% osmium tetroxide solution, 1 in 5 in dH$_2$O.

7. To each of the 5 min osmium tetroxide tubes add 2.5 μl of the 10 × osmium tetroxide buffer, and 15 μl of the diluted osmium tetroxide solution,[h] and place in a 37°C heating block for 5 min. To each of the 1 min osmium tetroxide tubes add 2.5 μl of the 10 × osmium tetroxide buffer, and 15 μl of the diluted osmium tetroxide solution, and place in a 37°C heating block for 1 min.

8. Stop the reactions by adding 200 μl of HOT stop buffer and 750 μl of ice-cold 100% ethanol. Invert the tubes and place immediately at –20°C for at least 30 min.

9. To each of the 0 min osmium tetroxide tubes add 200 μl of HOT stop buffer and 750 μl of ice-cold 100% ethanol. Add 2.5 μl of the 10 × osmium tetroxide buffer, and 15 μl of the diluted osmium tetroxide solution and invert the tubes, and place immediately at –20°C for at least 30 min.

10. Centrifuge the tubes for 15 min at 13 000 r.p.m. Remove the supernatant[i] and wash the pellets with 70% ethanol.

Protocol 1. *Continued*

11. Freshly dilute the piperidine 1 in 10 in dH$_2$O.

12. Add 50 μl of the diluted piperidine to the pellets and vortex for 10 sec.

13. Incubate tubes for 30 min in a 90°C heating block.

14. Chill the tubes on ice for 2 min,[j] then add 50 μl of 0.6 M NaAc pH 5.2 and 300 μl of ice-cold 100% ethanol, and 2.5 μl of glycogen[k] (20 mg/ml). Leave the tubes at –20°C for 30 min. Centrifuge the tubes for 15 min at 13000 r.p.m. Remove the supernatant and wash the pellets with 70% ethanol.

15. Count the dry pellets[b] in a β-counter and resuspend the pellets at 1000 c.p.m./2.5 μl in a 1 in 2 dilution of the loading dye.

D. *Electrophoresis of cleaved products*

1. Pour an 8% denaturing polyacrylamide sequencing gel.

2. Pre-run the gel for 1 h.

3. Denature the samples and an end-labelled DNA molecular weight marker (usually φX174 digested with *Hae*III) by heating at 95°C for 5 min and place immediately on ice.

4. Load 2.5 μl of the samples on the gel.[l]

5. Run the gel at 50 W until the bromophenol blue (BPB) dye reaches the bottom of the gel.

6. Dry the gel and place in a cassette with X-ray film.

7. Develop the film.

E. *Interpretation of results*

1. Compare the bands in each of the 20 min and 60 min hydroxylamine tracks with the zero time point for the hydroxylamine reaction for that heteroduplex. Compare the bands with hydroxylamine reactions for the homoduplex control. Any bands present in the patient's reactions that are not present in either the zero time point or homoduplex control are mismatch bands. A mismatch band in the hydroxylamine track indicates a C mismatch.

2. Compare the bands in each of the 1 min and 5 min osmium tetroxide tracks with the zero time point for the osmium tetroxide reaction for that heteroduplex. Compare the bands with osmium tetroxide reactions for the homoduplex control. Any bands present in the patient's osmium tetroxide reactions that are not present in either the zero time point or homoduplex control are mismatch bands. A mismatch band in the osmium tetroxide track indicates a T mismatch.

3. Determine the position of the mismatch by comparison with the end-labelled DNA molecular weight marker.

4. Sequence the appropriate segment[m] of the PCR product to identify the mutation.

[a] The patient PCR fragments must also be purified if both the normal and patient PCR fragments are to be labelled.

[b] Do not over-dry the pellet or it will be difficult to resuspend.

[c] If the volume of dH$_2$O required for a concentration of 1000 c.p.m./μl is greater than 1 ml, then a higher concentration can be used so that the probe can be resupended in the 1.5 ml micro-centifuge tube using a 1 ml Gilson.

[d] This may require leaving the DNA in the dH$_2$O at room temperature for 30 min and repeated attempts at resuspension with the 1 ml Gilson.

[e] Control homoduplexes are formed, as well as the heteroduplexes.

[f] This is to prevent the lid of the tube from opening while the tube is being boiled.

[g] For six chemical cleavage of mismatch reactions, for each heteroduplex, a minimum of 36 μl is needed. If there is less than 24 000 c.p.m. it may not be possible to do all six reactions.

[h] A yellow precipitate should appear.

[i] The hydroxylamine and osmium tetroxide wastes should be collected in separate containers for disposal.

[j] The tubes with the osmium tetroxide reactions should look slightly black. The tubes with the zero time point for the osmium tetroxide reactions will not be black.

[k] The glycogen helps precipitate the DNA and makes the pellet easier to resuspend in dye.

[l] Load the samples so that the six reactions for each patient are together. The six reactions for the homoduplex should be on the same gel so that the bands can be compared. Suggested order of the samples is 0, 1, 5 min osmium tetroxide then 0, 20, 60 min hydroxylamine for each heteroduplex.

[m] As both stands of the DNA have been 5' end-labelled, the mutation could be the determined distance in from either end of the PCR fragment.

Protocol 2. Single tube chemical cleavage of mismatch (STCCM) using a ^{32}P end-labelled probe[a]

Equipment and reagents
- See *Protocol 1*

A. *[32]P end-labelled probe preparation*

1. Follow *Protocol 1*.

B. *Heteroduplex formation*

1. As for *Protocol 1* except that only 250 ng of unlabelled DNA and 25 μl (25 000 c.p.m.) of probe are required.

C. *Chemical modification reactions*

1. For each heteroduplex aliquot 6 μl (6000 c.p.m.) into three separate 1.5 ml microcentrifuge tubes. Label each tube with the heteroduplex number and either 0 (zero control), S (short incubation), or L (long incubation).

Protocol 2. *Continued*

2. Perform all of the following manipluations in a fume-hood.

3. To each of the L tubes add 20 μl of the hydroxylamine solution and place in a 37°C heating block for 1 h. To each of the S tubes add 20 μl of the hydroxylamine solution and place in a 37°C heating block for 20 min.

4. Stop the reactions by adding 200 μl of HOT stop buffer and 750 μl of ice-cold 100% ethanol. Invert the tubes and place immediately at −20°C for at least 30 min.

5. To each of the 0 tubes add 200 μl of HOT stop buffer and 750 μl of ice-cold 100% ethanol. Add 20 μl of the hydroxylamine solution and invert the tubes, and place immediately at −20°C for at least 30 min.

6. Centrifuge the tubes for 15 min at 13 000 r.p.m. Remove the supernatant and wash the pellets with 70% ethanol.

7. Resuspend the pellets in 6 μl of dH$_2$O.

8. Freshly dilute the 4% osmium tetroxide solution, 1 in 5 in dH$_2$O.

9. To each of the L tubes add 2.5 μl of the 10 × osmium tetroxide buffer and 15 μl of the diluted osmium tetroxide solution, and place in a 37°C heating block for 5 min. To each of the S tubes add 2.5 μl of the 10 × osmium tetroxide buffer and 15 μl of the diluted osmium tetroxide solution, and place in a 37°C heating block for 1 min.

10. Stop the reactions by adding 200 μl of HOT stop buffer and 750 μl of ice-cold 100% ethanol. Invert the tubes and place immediately at −20°C for at least 30 min.

11. To each of the 0 tubes add 200 μl of HOT stop buffer and 750 μl of ice-cold 100% ethanol. Add 2.5 μl of the 10 × osmium tetroxide buffer and 15 μl of the diluted osmium tetroxide solution, invert the tubes, and place immediately at −20°C for at least 30 min.

12. Centrifuge the tubes for 15 min at 13 000 r.p.m. Remove the supernatant and wash the pellets with 70% ethanol.

13. Freshly dilute the piperidine 1 in 10 in dH$_2$O.

14. Add 50 μl of the diluted piperidine to the pellets and vortex for 10 sec.

15. Incubate tubes for 30 min in a 90°C heating block.

16. Chill the tubes on ice for 2 min, then add 50 μl of 0.6 M NaAc pH 5.2 and 300 μl of ice-cold 100% ethanol, and 2.5 μl of glycogen (20 mg/ml). Leave the tubes at −20°C for 30 min. Centrifuge the tube for 15 min at 13 000 r.p.m. Remove the supernatant and wash the pellet with 70% ethanol.

17. Count the dry pellets in a β-counter and resuspend the pellets at 1000 c.p.m./2.5 μl in a 1 in 2 dilution of the loading dye.

D. *Electrophoresis of cleaved products*

1. As for *Protocol 1* except that each heteroduplex only requires three tracks on the gel instead of six.

E. *Interpretation of results*

1. As for *Protocol 1* except that both hydroxylamine and osmium tetroxide reactions are in the same track. The type of mismatch can not be determined prior to sequencing.

---

[a] See relevant footnotes in *Protocol 1*.

2.1 Comments on the basic procedures

Uniformly labelled probes allow the detection of both segments of the cleaved strand of DNA. The size of the cleaved bands should add up to the full-length of the probe. This procedure, however, may result in complex patterns of cleaved bands if there is more than one mutation in a stretch of DNA being scanned. Therefore, end-labelled probes should be used instead of uniformly labelled ones in these circumstances. Of course the sequence variation relevant to this issue may include polymorphisms. Uniformly labelled probes also produce a higher background than end-labelled probes. This may obscure cleavage bands especially in the higher part of the gel.

T:G mismatches where the T is 3′ to a G are resistant to osmium tetroxide treatment (6, 18). If such a T is in the probe strand, the complementary heteroduplex has an A:C mismatch where C is in the target strand. Hydroxylamine treatment cleaves the mismatch but autoradiography does not reveal this event unless the target is labelled. This can be achieved by labelling both probe and target in one go or by inverting the labelling in successive experiments (e.g. first probe is labelled then target is labelled).

The above protocols are recommended for people setting-up the method. In normal practice several simplifications are possible:

(a) The time course (time 0, 1, and 2) of the chemical reaction is unnecessary. A single treatment time is sufficient and success of the chemical reaction is then controlled by cleavage of a heteroduplex containing a known appropriate mismatch.

(b) The homoduplex control can also be omitted, but the probe DNA exposed to chemical treatment provides the appropriate negative control while the probe unexposed to chemical treatment should be also examined in the gel to assess its quality and possible degradation.

Essential to the success of the procedure is the formation of labelled heteroduplexes in sufficient quantities to detect the cleavage products. The amount of label in the heteroduplex may be varied by altering the ratio of probe to target DNA, or the amount of total DNA used. When the amount of DNA is fixed, the amount of labelled heteroduplex can be increased by using a high ratio of target to probe (e.g. 10:1). Size of the wells in the gel plates and quality of the probe are important variables in the procedure. It should be noted that some mismatches may destabilize the DNA in such a way that modification and cleavage occurs additionally to a position adjacent to the mismatch. Such an effect is also responsible for cleavage of probe strands annealed to a target containing a single base or larger insertions. These insertions or deletions are always detected by the chemical mismatch detection method.

The above procedures do not require any specialist equipment.

3. Ultrafast chemical mismatch detection

At least a tenfold increase in the speed of mutation scanning can be obtained by using solid phase chemistry, fluorescent labelling, and automated fluorescent gel scanning. The protocols for this ultrafast mutation detection approach are based on the following principles.

3.1 Labelling

There are essentially two ways of fluorescently labelling PCR products to be visualized on the ABI Prism systems: end-labelling by incorporation of fluorescently tagged primers, or internal labelling using fluorescent dNTPs during PCR. Fluorescent primers can be made either using fluorescent phophoramidites (6-FAM, HEX, or TET), or by synthesizing the oligonucleotide with an aminolink group at the 5′ end and then reacting this link with the NHS-ester dye (e.g. TAMRA and ROX). The former method is much more efficient and, if HPLC or column purification is used it ensures that virtually 100% of the primer, and hence PCR product, is labelled. Fluorescent dNTPs (i.e. dUTP or dCTP) have been made available more recently and allow PCR internal labelling which may give more fluorophores per length of DNA, and thus allows greater sensitivity so that less DNA can be used.

3.2 Solid phase

The standard procedure of chemical cleavage of mismatch is rather laborious as it involves at least three ethanol precipitations. This limits the number of tubes that can reliably be handled by one person in one day to about 24 (12 with hydroxylamine and 12 with osmium tetroxide). Streptavidin-coated magnetic beads are now available from several suppliers (Dynal, Promega) and will readily bind biotinylated DNA in high salt. Thus, if 5′ biotinylated primers are used to generate a PCR product, and hybridized to a probe, the resultant heteroduplexes can be bound to the magnetic beads. All subsequent handling simply involves placing the tube (or microtitre plate) on a magnet while removing the liquid, then resuspending in the new solution (e.g. hydroxylamine). This procedure minimizes the volume of chemical solution to be disposed of (i.e. no ethanol washes) and greatly speeds up the entire protocol. The piperidine treatment cleaves, denatures and, at least in part, releases the DNA from the beads so that the eluant must be retained for gel loading. For this reason, a small volume of piperidine mixed with the formamide/loading dye is used so that after the 90°C incubation, the supernatant can be loaded directly on a gel. Alternatively, an ethanol precipitation at this stage will remove traces of piperidine and also the low molecular weight fluorescent artefacts usually seen after osmium treatment of internally labelled products. The process is shown schematically for both end and internal labelling in *Figures 2* and *3*.

Protocol 3.   Solid phase chemical cleavage with end-labelling

Equipment and reagents

- ABI 377 Prism or 373 DNA sequencer
- Microtitre plate magnet and/or microcentrifuge tube magnet
- Streptavidin-coated magnetic beads (Promega)
- 10 × hybridization buffer: 3 M sodium chloride, 100 mM Tris–HCl pH 8
- 2 × binding buffer: 2 M sodium chloride, 0.4% Tween 20, 10 mM Tris–HCl pH 8, 0.1 mM EDTA

- Hydroxylamine solution: 4 M hydroxylamine hydrochloride (Aldrich), 2.3 M diethylamine pH 6
- Osmium solution: 0.4% osmium tetroxide, 2% pyridine (Aldrich) (osmium tetroxide can be purchased as a 4% solution and stored at 4°C for a maximum of three months)
- Piperidine/dyes: 1 M piperidine in formamide/dextran blue loading dye

Method

1. Fluorescently labelled primers are used to amplify the DNA of interest from a normal individual (or cDNA) to generate the probe DNA.

2. Gel purify probe DNA, for example using GeneClean (Bio 101), resuspend in one-fifth of the original volume, and store in the dark at 4°C.

3. Amplify the corresponding target sequences using biotinylated primers at a final concentration of about 1 ng/µl.[a] Up to four different PCRs can be multiplexed if desired.[b]

4. Add 5 µl of each purified end-labelled probe to 20 µl of the four multiplexed non-purified targets. Add 0.1 vol. of 10 × hybridization buffer and incubate at 95°C for 5 min, followed by 65°C for 1 h.[c]

5. Prepare the streptavidin magnetic beads by taking a volume that is equal to the total volume of biotinylated targets (i.e. number of tubes × 20 µl) and washing twice in 2 × binding buffer, before resuspending in a volume equal to the total volume of hybrids generated.[d]

Protocol 3. *Continued*

6. Add an equal volume of the washed beads to the hybrids. Mix and leave at room temperature for about 15 min.

7. Remove and discard supernatant while the tubes/plate are on the magnet.

8. Add 20 μl of hydroxylamine solution or osmium solution to the beads and resuspend. Incubate for 2 h at 37 °C for hydroxylamine (or 15 min at room temperature for osmium).

9. Place on magnet and remove supernatant. Resuspend beads in 5 μl of piperidine solution and heat to 90 °C for 30 min.

10. Snap chill on ice/water before loading supernatant on a 6% polyacrylamide gel on an ABI 373 DNA sequencer.

11. Run at 41 W for 14 h on standard sequencing plates for fragments up to 1.8 kb. Alternatively, run on an ABI 377 on 12 cm well-to-read plates for 6 h under module GS12–2400A.

[a] Low concentrations of biotinylated primers must be used to ensure that excess unused primers do not compete with the PCR product for the streptavidin beads.
[b] In our experience the ROX labelled primers are the least sensitive and only work well with small PCR products.
[c] It is sometimes convenient to leave at 65 °C overnight.
[d] It may be necessary to titre the beads against the PCR product to ensure most of it is binding.

Protocol 4. Solid phase chemical cleavage of mismatch with internal labelling

1. Prepare internally labelled probes by amplifying the desired section from normal DNA (or cDNA) using standard primers plus one of the three available fluorescent dUTPs (or dCTPs) from ABI: R6G, R110, or TAMRA at the manufacurer's recommended concentration.

2. Gel purify the products, for example using GeneClean, and resuspend in the same original volume of TE, and store in the dark at 4 °C.

3. Amplify the target DNA using biotinylated primers[a] at a final concentration of about 1 ng/μl, plus the same fluorescent dNTP as used in the corresponding probe section. 10 μl PCRs are sufficient.

4. Add 1 μl of each differently labelled section (up to three) plus 1 μl of each probe. Add 3 μl of 10 × hybridization buffer and make up to 30 μl.

5. Add a drop of oil and heat to 95 °C for 5 min, followed by 65 °C for 1 h.

6. Add 30 μl of pre-washed magnetic beads (see *Protocol 3*, step 5) and leave at room temperature for 15 min.

7. Remove supernatant and proceed as in *Protocol 3*, steps 7–9.

[a] Biotinylated primers are expensive. Therefore, if only a few reactions are to be carried out, it may be desirable to combine internal fluorescent labelling with ethanol preciptations (see *Protocol 2*).

5: Cleavage of mismatched bases using chemical reagents

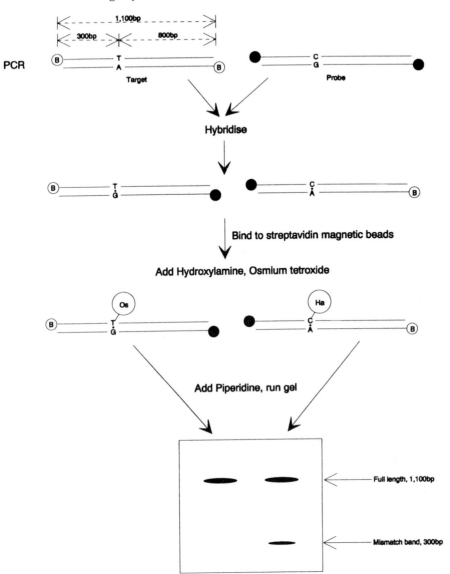

Figure 2. Fluorescent end-labelling and solid phase chemical cleavage of mismatch. In the figure, only one fragment of a possible three or four multiplex is shown. The B represents the biotinylated primers, while the shaded circle represents the fluorescent primers. The target is shown with a C to T mutation at a position 300 bp from the 5′ end of an 1100 bp PCR product. Only the labelled products are seen on the fluorescent gel.

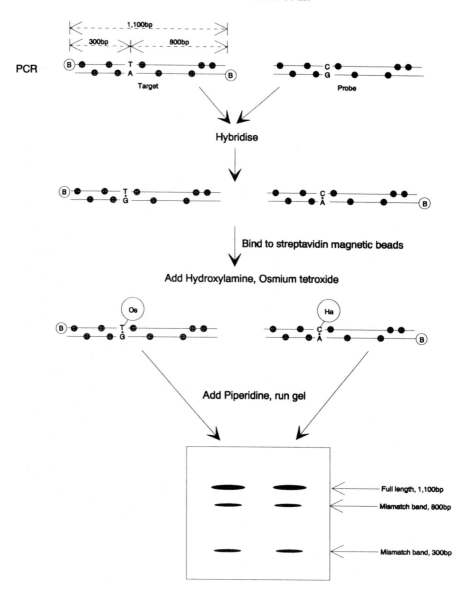

Figure 3. Fluorescent uniform labelling and solid phase chemical cleavage of mismatch. The same mutation and symbols are used as for *Figure 2*. Compare with *Figure 2* to see difference in bands expected on the gel.

94

3.3 Comments

The two methods of fluorescent labelling presented here have slightly different advantages and disadvantages. End-labelling, by using fluorescent
primers, allows up to four segments to be mismatched per gel track and also
allows multiplexing of the PCR. Thus, if four 1.5 kb fragments are multiplexed, it is possible to screen 6 kb of DNA per gel track. Alternatively, if

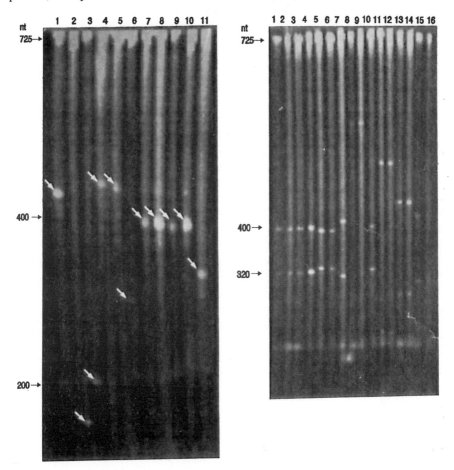

Figure 4. Gel images of mismatch cleavage products obtained by *Genescan* software
(ABI) after electrophoresis of denatured hybrids exposed to the solid phase mismatch
procedure. (A) Known mutations in exon h of factor IX revealed by using a probe generated with HEX labelled primers after hydroxylamine modification. Each track shows at
the top the full-length band (725 nt), and one mismatch band (arrows), of a size appropriate to the known mutation. (B) Mismatch bands due to exon h mutations in patients and
ascendant relatives. Hybrids formed by internally labelled (R6G dUTP) probe and target
DNA were treated with hydroxylamine. Two bands are seen for each mutation adding up
to the full-length probe (725 nt). Reproduced from ref. 12 with permission.

different fluorophores are used on the 5′ and 3′ primers, then more precise positional information can be obtained at the expense of screening less fragments simultaneously. This may be useful if only one or two PCRs (or RT-PCRs) are required to cover a region of interest.

Internal labelling allows up to three PCR products to be mixed together and mismatched (as only three fluorescent dNTPs are available). However, multiplexing is obviously impossible during the PCR stage. The main advantage of using internal label is that both target and probe can be labelled thus usually providing two chances of detecting a point mutation. The exceptions are C:C and T:T mismatches, whose complementary hybrids (G:G and A:A) will not react with either chemical, and the T:G mismatch where the T is 3′ to a G because, as mentioned earlier, this is resistant to chemical treatment. The alternative hybrid with the C:A mismatch will reveal the presence of a sequence change after hydroxylamine treatment. Since biotin labelling and streptavidin beads binding allow the selection of one of the two DNAs used to form heteroduplexes, this procedure allows greater flexibility in the ratio of probe to target.

The solid chemistry fluorescent mismatch procedure was initially developed using haemophilia B as a model. The defective gene, factor IX, consists of eight exons that comprise 1.4 kb of coding sequence. These were amplified in seven sections from DNA (exons b and c in one fragment) and then sorted into two multiplexes. For end-labelled PCR, one multiplex contained four fragments and the other three. For internal labelling two multiplexes of three fragments each were used (exon e being sequenced instead). In this way the entire coding sequence could be screened by fluorescent chemical cleavage in two multiplexes with one chemical (i.e. four gel lanes when both chemicals were used). An example gel picture is shown in *Figure 4*. This number could be reduced by using the chemicals sequentially as described in *Protocol 2* and ref. 15.

The technique has since been successfully applied to the BRCA1 gene (19) and factor VIII (20) using RT-PCR on segments up to 1.3 kb and DNA PCR up to 1.8 kb.

References

1. Cotton, R. G. H., Rodrigues, N. R., and Campbell, R. D. (1988). *Proc. Natl. Acad. Sci. USA*, **85**, 4397.
2. Montandon, A. J., Green, P. M., Giannelli, F., and Bentley, D. R. (1989). *Nucleic Acids Res.*, **17**, 3347.
3. Dahl, H-H. M., Lamande, S. R., Cotton, R. G. H., and Bateman, J. F. (1989). *Anal. Biochem.*, **183**, 263.
4. Naylor, J. A., Green, P. M., Montandon, A. J., Rizza, C. R., and Giannelli, F. (1991). *Lancet*, **337**, 635.
5. Roberts, R. G., Bobrow, M., and Bentley, D. R. (1992). *Proc. Natl. Acad. Sci. USA*, **89**, 2331.

6. Forrest, S. M., Dahl, H. H., Howells, D. W., Dianzani, I., and Cotton, R. G. H. (1991). *Am. J. Hum. Genet.*, **49**, 175.
7. Dianzani, I., Forrest, S. M., Camaschella, C., Gottardi, E., and Cotton, R. G. H. (1991). *Am. J. Hum. Genet.*, **48**, 423.
8. Howells, D. W., Forrest, S. M., Dahl, H-H. M., and Cotton, R. G. H. (1990). *Am. J. Hum. Genet.*, **47**, 279.
9. Saleeba, J. A. and Cotton, R. G. H. (1991). *Nucleic Acids Res.*, **19**, 1712.
10. Saleeba, J. A., Ramus, S. J., and Cotton, R. G. H. (1992). *Hum. Mutat.*, **1**, 63.
11. Wurst, H. and Pohl, F. M. (1991). *Proc. Natl. Acad. Sci. USA*, **88**, 9909.
12. Rowley, G., Saad, S., Giannelli, F., and Green, P. M. (1995). *Genomics*, **30**, 574.
13. Haris, I. I., Green, P. M., Bentley, D. R., and Giannelli, F. (1994). *PCR Methods Appl.*, **3**, 268.
14. Verpy, E., Biasotto, M., Meo, T., and Tosi, M. (1994). *Proc. Natl. Acad. Sci. USA*, **91**, 1873.
15. Ramus, S. J. and Cotton, R. G. H. (1996). *BioTechniques*, **21**, 216.
16. Gogos, J. A., Karayiorgou, M., Aburatani, H., and Kafatos, F. C. (1990). *Nucleic Acids Res.*, **18**, 6807.
17. Hansen, L. L., Justesen, J., and Kruse, T. A. (1996). *Hum. Mutat.*, **7**, 256.
18. Cotton, R. G. H., Dahl, H-H. M., Forest, S. M., Howells, D. W., Ramus, S. J., Bishop, R. E., *et al.* (1993). *DNA Cell Biol.*, **12**, 945.
19. Greenman, J., Mohamed, S., Ellis, D., Watts, S., Scott, G., Barnes, D., *et al.* (1997). *Genes, Chromosomes and Cancer*, in press.
20. Waseem, N. H., Bagnall, R., Green, P. M., and Gianelli, F. (1997). *Thromb. Haemostasis*, Supplement, 228.

6

Mutation detection using T4 endonuclease VII

RIMA YOUIL

1. Introduction

The ideal method for analysis of genes implicated in human disease would be a simple, easy to perform procedure that utilizes standard technology available in most laboratories. It should offer high throughput capacity even when performed manually. The procedure should also lend itself to automation to allow for increased throughput, particularly for laboratories that routinely handle large numbers of samples. The ability of the scanning technique to localize the point of mutation would also be highly advantageous.

This chapter describes a new method known as enzyme mismatch cleavage (EMC) that is now approaching this ideal. EMC utilizes the enzyme, T4 endonuclease VII (Endo VII), for the purpose of mutation detection. Endo VII has been shown to detect minor perturbations in heteroduplex DNA resulting from mutations or polymorphisms, with high sensitivity. It is capable of scanning large fragments of DNA and localizing the position of mutations. In addition to the simplicity of the EMC method, the robust nature of the enzyme allows for conversion of this technology to automation.

2. The biology of Endo VII

2.1 The role of Endo VII *in vivo*

Endo VII is the product of gene 49 of the bacteriophage T4 (1). This enzyme has been well characterized in its role as a 'resolvase'. *In vivo*, Endo VII has a multitude of roles—being responsible for the resolution of complex replication intermediates in T4 reproduction as well as being involved in DNA repair (2–5).

2.2 Characterization of Endo VII

Endo VII has been purified to homogeneity (6, 7). SDS–polyacrylamide gels show that purified Endo VII has an apparent molecular mass of 17.8 kDa (6). It has been determined that Endo VII functions as a homodimer (6).

Endo VII is a 157 amino acid protein (8, 9). There is a zinc-binding domain towards the N terminal end of the gene, predicting that the gene product has a DNA binding role.

The natural substrate for Endo VII is branched DNA. Much of the characterization of Endo VII has utilized rapidly sedimenting DNA (rsDNA) as substrate. Endo VII has been shown to have a broad temperature and pH range with maximal activity obtained at 37°C in Tris–HCl buffer at pH 8.6 (10). The enzyme requires $MgCl_2$ for cleavage with optimum concentration at 10 mM (10). It has been shown that in the absence of $MgCl_2$, Endo VII will only bind to its substrate (11).

2.3 Action of Endo VII on heteroduplex DNA

Subsequent to the demonstration that Endo VII can resolve Holliday junctions *in vitro*, it has been shown that Endo VII can recognize and cleave a broad spectrum of DNA structures (12). These structures include cruciforms (13), Y junctions (14), semi-Y junctions, single-strand overhangs, nicks and gaps (15), heteroduplex loops (16), terminal loops (17), bulky adducts (18), cisplatin adducts (19), and curved DNA (20).

In an effort to identify the smallest distortion to the helical structure that Endo VII can recognize, Solaro *et al.* (2) tested various single base pair mismatches as well as single base pair loops in short oligonucleotides. These authors observed that Endo VII was able to bind and cleave at every single mismatch and loop type—albeit with varying efficiencies. The less thermostable mismatches (C-containing mismatches) were cleaved more readily than the more thermostable mismatches (G-containing mismatches).

Solaro *et al.* (2) have shown that Endo VII introduces double-stranded cleavage on the 3' side and within 6 bp of the distortion. Cleavage of duplex DNA results in 3' OH and 5' PO_4^- ends.

Heteroduplex loops formed between normal DNA and mutant DNA containing an insertion or deletion, are detected by Endo VII with high efficiency (2). Resolution of loop structures occurs by cleavage at the double-stranded portion of DNA flanking the loop. Similar to the cleavage of heteroduplexes, cleavage is within six base pairs from the base of the loop (16). The size of the loop will not affect cleavage. The loop structure itself is not a substrate for Endo VII (16) so long as the loop does not contain any complementary regions that would initiate secondary structure formation. Birkenkamp and Kemper (21) have observed that secondary structures within large loops are recognized by Endo VII. These authors also observed that Endo VII is capable of efficiently repairing heteroduplex loops in a manner similar to its repair of mismatches in duplex DNA.

Unlike restriction endonucleases, Endo VII may choose to cleave the DNA at more than one site of the mismatch, normally choosing a major cleavage site and perhaps one or several minor and subminor sites (11, 15, 16).

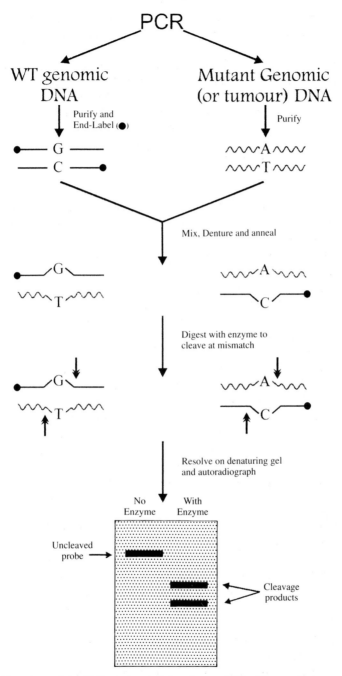

Figure 1. Flowchart of the EMC protocol. Taken from ref. 24 with permission.

3. Use of Endo VII for mutation detection

3.1 Enzyme mismatch cleavage

The EMC method utilizes Endo VII to cleave at mismatches present in heteroduplex molecules formed by mixing, denaturing, and annealing reference and mutant DNA (22, 23). *Figure 1* shows an outline of the EMC strategy.

3.2 Amplification of reference and target DNA

The source of the DNA may be plasmid, genomic, cDNA, or mitochondrial DNA. PCR amplification reactions must be optimized to produce good yield of the correctly sized product with no extraneous bands. It is important to produce 'clean' PCR products to avoid purification of target DNA samples. The reference sample, on the other hand, requires purification for removal of excess PCR primers to ensure high specific labelling of the reference PCR product.

The amount of reference DNA chosen for labelling will depend on the number of samples to be scanned. The reference DNA can be prepared in bulk to support any number of assays. The reference DNA must be purified prior to performing the end-labelling reaction. Purification of reference DNA from the PCR reaction can be performed by gel purification (Qiagen, CA) or by using a Centricon G100 unit (Amicon, MA) following the manufacturer's instructions. The volume of reference DNA PCR reaction to be purified depends on the number of assays to be performed. It is recommended that a 1 ml volume of PCR amplified reference DNA be purified. Label only that amount of DNA required. The remaining DNA can be stored at –20°C. Probe (prepared with fresh label) can be used over a period of one month.

Protocol 1. Preparation of reference DNA probe

Equipment and reagents

- Mini-gel apparatus
- 10 × TBE: 89 mM Tris base, 89 mM boric acid, 2 mM EDTA
- T4 polynucleotide kinase (Boehringer Mannheim)
- [γ-^{32}P]ATP (6000 Ci/mmol)
- 10 × T4 polynucleotide kinase buffer (Boehringer Mannheim)
- TE buffer: 10 mM Tris–HCl, 1 mM EDTA pH 8.0
- G50 spun column (Pharmacia)

Method

1. Quantitate the amount of DNA present in 1 μl of the purifed reference DNA by electrophoresis on a 1% mini agarose gel with 0.5 × TBE buffer. Following ethidium bromide staining, estimate DNA concentration of the sample by comparing band intensity against a known molecular weight marker run on an adjacent lane of the same gel.

---

2. Label 5 picomole of purified reference DNA as follows:

- 8.5 μl (5 picomole) DNA
- 2.0 μl 10 × T4 polynucleotide kinase buffer
- 1.5 μl T4 polynucleotide kinase (10 U/μl)
- 8.0 μl [γ-^{32}P]ATP (6000 Ci/mmol)

Incubate tube for 45 min at 37 °C.

3. Add TE buffer to a total volume of 50 μl.

4. Remove unincorporated label using a G50 spun column.

5. Determine the specific activity of probe as c.p.m./picomole. Aim for no less than ~ 2 × 10^6 c.p.m./picomole (or 200 000 c.p.m./μl) probe.

3.3 Formation of heteroduplexes

The quality of heteroduplex DNA molecules formed by hybridization of reference DNA with a test DNA is critical to the sensitivity and signal-to-noise ratio obtained in any mismatch detection assay. It has been well established that Endo VII has a high recognition for DNA secondary structure. Mismatch detection assays depend on preferential recognition of mismatch DNA above any other sequence-induced structural distortion. Therefore, optimal conditions must be devised to allow for generation of heteroduplex molecules.

To distinguish signal from noise, homoduplex controls must also be prepared. This requires hybridization of probe with an excess of unlabelled reference DNA. This DNA must be the same reference DNA used in the preparation of the probe to minimize complexity due to polymorphisms.

Protocol 2. The annealing reaction

Reagents

- 2 × annealing buffer: 1.2 M NaCl, 12 mM Tris–HCl pH 7.5, 14 mM MgCl$_2$
- TE buffer (see *Protocol 1*)

Method

1. Dilute the reference probe to ~ 200 000 c.p.m./μl with TE. Mix probe and reference or test DNA together in the presence of 2 × annealing buffer as follows.

2. (a) For the reference control sample add the following:
   2 μl of labelled reference probe (~ 400 000 c.p.m.)
   10 μl of reference DNA PCR sample[a] (0.1–1.0 picomole)
   1.5 μl of 2 × annealing buffer
   1.5 μl of dH$_2$O

Protocol 2. *Continued*

    (b) For the test sample add the following:
        2 μl of labelled reference probe (~ 400 000 c.p.m.)
        10 μl of each test DNA[b] PCR sample[a] (0.1–1.0 picomole)
        1.5 μl of 2 × annealing buffer
        1.5 μl of dH$_2$O

3. Heat samples to 95°C for 5 min. Remove tubes immediately to a 65°C heating block and let tubes rest for 1 h.

4. Incubate tubes for a further hour at room temperature before precipitating with sodium acetate/ethanol.

5. Wash pellet with 500 μl of 70% ice-cold ethanol and thoroughly resuspend dried pellet in 8 μl of dH$_2$O.

[a] Test and reference samples used in the annealing reaction can be taken directly from the PCR reaction without purification.
[b] Test DNA may be either homozygous or heterozygous for the mutation.

To form heteroduplexes containing imperfectly paired bases, there must be two variations of a DNA molecule present in the annealing reaction. DNA samples heterozygous for a mutation will contain a 'normal' and a 'mutant' version of the allele. Such a sample will require only melting and annealing of that DNA sample. Theoretically this will result in the formation of four different DNA molecules. Two of the molecular species will be homoduplexes (formed by regeneration of the perfectly matched normal and mutant alleles). The remaining two will be heteroduplex molecules each carrying one normal and one mutant DNA strand. Any base pair difference between the two strands will result in a 'bulge' in the DNA structure at the exact site of the base change. This 'bulge' is referred to as a mismatch and results from the pairing of non-complementary bases.

For a homozygous DNA sample containing two copies of the mutant allele, heteroduplexes can only be formed by the addition of DNA homozygous for the normal allele to the mutant DNA sample.

Table 1. The four types of single base pair mutation combinations possible after heteroduplex formation between wild-type and mutant DNA

	Base change	Mismatch set
Type 1	A→G or G→A	A:C and G:T
	C→T or T→C	
Type 2	G→T or T→G	G:A and T:C
	A→C or C→A	
Type 3	A→T or T→A	A:A and T:T
Type 4	C→G or G→C	C:C and G:G

In either situation, the type of mismatch that results will depend on the base difference between the normal and mutant allele. There are 12 different base changes that can occur in DNA and these can result in any of eight different mismatches. These eight mismatches fall into four mismatch sets (see *Table 1*).

3.4 Endo VII digestion of heteroduplex DNA

Following the annealing reaction, the DNA samples are then digested with Endo VII (500 U/μl). As Endo VII may be supplied at higher concentrations, it may be diluted using enzyme dilution buffer: 10 mM Tris pH 8.0, 5 mM DTT, 20% glycerol (v/v), 100 mg/ml BSA.

Protocol 3. The Endo VII cleavage reaction

Reagents

- 10 × reaction buffer: 500 mM Tris–HCl pH 8.0, 100 mM MgCl$_2$, 100 mM DTT, 1 mg/ml BSA
- T4 Endo VII[a] (1000 U/μl)

- Formamide loading buffer: 95% high quality, deionized formamide, 20 mM EDTA, 0.05% xylene cyanol, 0.05% orange G

Method

1. Set-up the following reactions:
   - 8 μl annealed DNA (from *Protocol 2*)
   - 1 μl 10 × reaction buffer
   - 1 μl T4 Endo VII (1000 U/μl)
2. Tubes are incubated at 37 °C for 30 min.
3. Following incubation add 10 μl of formamide loading buffer to each tube. The tubes are heated to 95 °C for 5 min and immediately placed on ice ready for fragment analysis.

[a] The T4 Endo VII enzyme will be supplied commercially as part of the 'EMD' kit (see Section 7.).

3.5 Analysis of cleavage products

The digestion products from each of the reactions are denatured and resolved on a denaturing polyacrylamide gel. The presence of unique cleavage products present in the test lanes indicate the site of mutation or polymorphism. The location can be determined by sizing the products to a molecular size marker present on the same gel.

Protocol 4. Resolution of digestion products

Equipment and reagents

- Standard sequencing apparatus
- 8% polyacrylamide, 7 M urea mix

- TBE
- X-ray film

105

Protocol 4. *Continued*

Method

1. Prepare 0.4 mm sequencing gel using a well-forming comb.

2. Pre-warm the 8% polyacrylamide, 7 M urea denaturing gel in 0.5 × TBE buffer (or 6% Long Ranger™ polyacrylamide: 8.3 M urea in 0.6 × TBE buffer) to 55°C. Long Ranger™ gel solution can be purchased from FMC BioProducts, Rockland, ME.

3. Load 5 μl of each sample (this should amount to ~ 80 000–100 000 c.p.m. per sample). Load a radioactively labelled DNA molecular size marker in one lane of the gel.

4. Electrophorese samples for 1–1.25 h at 45 W, until the orange G dye reaches the bottom of the gel. This will be sufficient to resolve fragments of up to 1.5 kb.

5. The gel is then dried down onto blotting paper and exposed to X-ray film (Kodak) overnight at room temperature or 4 h at –70°C with intensifying screens.

If the reference and test DNA sequences have no differences between them then all duplex molecules will be perfectly matched. Any base pair discrepancies between the test DNA and the reference DNA will form a 'bulge' in the helix where the bases fail to match.

It is this bulge or distortion in the reference DNA structure that the Endo VII is able to recognize and subsequently cleave. Only the strand containing the radiolabel will be detected by autoradiography. Cleavage of the end-labelled reference DNA strands in either of the heteroduplexes will result in a DNA strand that is shorter than the uncleaved substrate.

Following resolution of the digestion products on a denaturing polyacrylamide gel, the cleavage products are detected by exposure of the gel to an X-ray film. The EMC autoradiograms are then analysed by comparing each test sample lane with the reference DNA lane. The presence of any extra bands in the test lane, when compared with the reference, signifies a positive result. The size of the band (when compared to radiolabelled DNA molecular weight size markers run on an adjacent lane on the same gel) will indicate the approximate position at which the base pair discrepancy occurs. However, it is not possible to predict which end of the molecule (5' or 3') the mutation falls. Sequencing at both ends of the molecule within the region indicated by EMC will need to be performed in order to define the base change.

The action of the enzyme on sequence-induced distortions can generate a background pattern of 'non-specific' cleavage. That is, products that do not result from specific cleavage of a mismatch or loop (23). Therefore, only bands present in the test that are NOT present in the homoduplex control lane signify a cleavage product resulting from cleavage of a mismatch or loop.

4. Fluorescent EMC

If preferred, the reference DNA may be labelled by fluorescence (*Figure 1*). This is performed simply by incorporating fluorescently labelled primers in the PCR reaction when amplifying reference DNA. The test samples are amplified with unmodified primers. The EMC procedure is similar to the radioactive method, except that fragments are analysed on an automated sequencer or by capillary electrophoresis.

Fluorescent EMC can also provide one with additional information as to the exact location of the mutation. Using differently labelled 5′ and 3′ primers, one is able to identify whether a mutation is present 5′ or 3′ of the scanned DNA fragment. This can drastically reduce the time and effort required to define the position of any mutation detected.

5. Factors to consider in the EMC assay

5.1 Incubation time and temperature

It has been demonstrated (Avitech Diagnostics, Inc., personal communication) that Endo VII works over a broad time and temperature range. Optimum condition for incubation is 30 min at 37 °C.

5.2 Applicability of EMC to various genes and DNA types

Endo VII has been used on a variety of genes including PDH E1 α subunit, 21-hydroxylase gene, α1 antitrypsin, β-globin, rhodopsin, DHPR, mouse mottled Menkes gene (23), mouse β-globin promoter (22), p53 (24), HLFMO3 (unpublished data), holocarboxylase synthetase gene (unpublished data), *M. tuberculosis rho* β gene (25), HNPCC (25), and BRCA1 (26).

5.3 Sensitivity

Sensitivity is of paramount importance in any mutation detection method. The detection rate for Endo VII has been estimated to be in the range of > 98% (22, 23). In a study of 81 known mutations (22), all 81 samples were positively identified. Cleavage detection varied from strong to weak cleavage. Mashal *et al.* (27) reported 3/14 mutations not detectable by Endo VII. One of the mutations missed was the G551D cystic fibrosis mutation. In a separate study, Avitech Diagnostics, Inc. has shown 99% detection in a study of 99 characterized mutations (personal communications). The G551D mutation was included in this study and was positively identified (personal communication, Avitech Diagnostics Inc.).

For the purpose of detecting a mutation, the enzyme needs only to cleave one of the two heteroduplexes present in each assay. Youil *et al.* (23) observed that some mutations resulted in cleavage of both mismatches (represented by the presence of two cleavage products from an end-labelled probe) and some

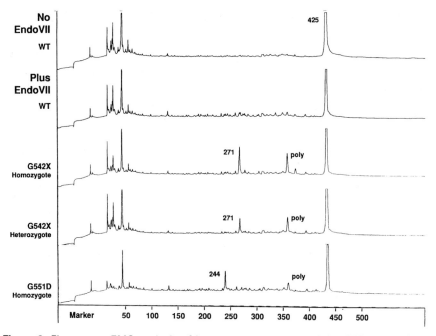

Figure 2. Fluorescent EMC analysis of heterozygous mutants of the *CFTR* gene. G542X and G551D and a silent polymorphism within exon 11 of *CFTR* were analysed on the ALF automated sequencer (Pharmacia Biotechnology). The G542X mutation was analysed in homozygous and heterozygous form. WT: wild-type sequence of exon 11 of the cystic fibrosis gene. The polymorphism is homozygous in both G542X mutants tested. However, G551D carries the polymorphism in the heterozygous form. There is a corresponding reduction in the peak height of the polymorphism comparing the homozygous with the heterozygous state. EMC was performed using manufacturer's (Avitech Diagnostics., Inc.) instructions. Partial heteroduplex digestion with Endo VII enables visualization of all base changes (particularly important for PCR amplicons that have multiple changes). Courtesy of Avitech Diagnostics., Inc.

resulted in only one cleavage product band. It is therefore essential that both heteroduplexes are labelled (at least one strand of each heteroduplex) to allow for increased chance of detection.

5.4 DNA fragment length

One of the benefits of Endo VII is its ability to detect mutations in large DNA fragments. It has been reported that Endo VII can detect single base pair mutations in 1377 bp fragments (23). It has since been shown that Endo VII can detect mutations in fragments greater than 2000 bp (personal communications, Avitech Diagnostics, Inc.). The ability to scan long fragments along with the ability to localize the mutation makes EMC highly advantageous, increasing the rapidity with which samples can be processed and analysed.

5.5 Detection of multiple mutations within a single DNA fragment

Highly polymorphic genes can be problematic for some mutation detection methods. This can be of particular concern when studying genetic diversity in micro-organisms for example. Ideally, a mutation detection method should be capable of identifying all base changes within such amplicons. Data on Endo VII accumulated over the last few years have indicated that this enzyme is capable of detecting multiple mutations within a single-strand of DNA (21–23) (*Figure 2*). Birkenkamp and Kemper (21) have also shown that Endo VII is capable of discriminating between a loop structure and a T:T mismatch arranged in tandem on the same molecule, six nucleotides from one another.

Endo VII identifies mutations even when they exist within high GC-containing sequences. A region of the *rho* β gene of *M. tuberculosis* (65.4% GC-rich) was successfully scanned for mutations (25, unpublished data).

5.6 Stability of Endo VII

Endo VII is stable at high protein concentrations when stored at –20°C (or –70°C for long-term storage) in buffer containing 10 mM Tris–HCl pH 7.5, 0.1 mM glutathione, and 50% (v/v) glycerol (6).

6. Application to solid phase and non-radioactive format

The EMC strategy may readily be adapted to solid phase (28) (*Figure 3*). It has been proposed that the background resulting from non-specific cleavage may be largely reduced if excess probe is removed from the annealing mixture (28). Using biotinylated primers to amplify test DNA sequences and performing the heteroduplex reaction with radiolabelled or fluoresceinated probe will allow isolation of heteroduplex sequences. Addition of magnetic streptavidin-coated beads to the annealed DNA, enables heteroduplex molecules (each carrying one radiolabelled strand and one biotinylated strand) to be separated from the reference probe. This step essentially selects for heteroduplex molecules which will be radioactively labelled. Non-labelled homoduplex test DNA will also be present, however, this will be invisible to detection by autoradiography.

Comparison between the two strategies have shown that there is marked improvement to the background. However, recent improvements to EMC have shown that similar results are achievable without the use of the solid phase system (Avitech Diagnostics, Inc., personal communication).

EMC can be easily adaptable to fluorescent-based detection systems. EMC cleavage products can be resolved on automated sequencing systems. This has proven to be quite a sensitive approach and capable of detecting multiple mutations within a single-strand of DNA (*Figure 2*).

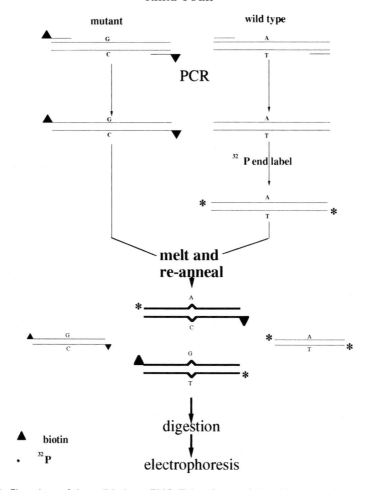

Figure 3. Flowchart of the solid phase EMC. Taken from ref. 28 with permission.

7. EMD™: a commercial kit

The EMC method has been further improved upon and has been renamed Enzymatic Mutation Detection™ (EMD™). The EMD method (developed by Avitech Diagnostics, Inc., PA) is a simple four step assay that offers high sensitivity and high signal-to-noise ratio. The kit, containing enzyme, buffers, control DNA, and protocol, should be available commercially by mid-1998. The new four step assay developed by Avitech Diagnostics eliminates the need for purifying or quantifying the test samples and reduces the annealing step to 10 minutes. The protocol provided with the EMD™ kit is applicable to all genes and DNA sources without any further optimization.

8. Future prospects

There are various reasons why researchers study sequence variability. The study of inherited diseases, epidemiological studies, candidate gene studies, phenotype cloning, or screening of site-directed mutagenized transcripts are all applications that require reliable mutation detection methods. Common to all of these applications is the need for a sensitive and rapid approach to analyses.

In the relatively short time that Endo VII has been proposed as a suitable candidate for the study for mutations, there have been intense efforts made towards defining the full potential of this enzyme on linear DNA substrates. In addition to assessing the capacity of the enzyme to detect all mismatches in different sequence contexts, efforts are being made to address specific areas of need.

The conversion of EMC to a non-radioactive mode is extremely important. The understandable concern over the use of radioactivity has steered mutation detection strategies toward other detection means that maintain high sensitivity. Fluorescent dyes are the primary choice since they offer the sensitivity and the capacity for automation. Significant efforts towards automation of EMC is presently underway (Avitech Diagnostics, Inc., personal communication).

Acknowledgements

Dr Christopher D. Earl and Ms Anne L. Bailey are thanked for their review of this manuscript. Avitech Diagnostics, Inc. is also thanked for allowing publication of their work (*Figure 2*).

References

1. Mizuuchi, K., Kemper, B., Hays, J., and Weisberg, R. A. (1982). *Cell*, **29**, 357.
2. Solaro, P. C., Birkenkamp, K., Pfeiffer, P., and Kemper, B. (1993). *J. Mol. Biol.*, **230**, 868.
3. Minagawa, T. and Ryo, T. (1978). *Virology*, **91**, 222.
4. Mosig, G. (1987). *Annu. Rev. Genet.*, **21**, 347.
5. Mosig, G., Luder, A., Ernst, A., and Canan, N. (1991). *New Biol.*, **3**, 1195.
6. Kosak, H. G. and Kemper, B. (1990). *Eur. J. Biochem.*, **194**, 779.
7. Golz, S., Birkenbihl, R. P., and Kemper, B. (1995). *DNA Res.*, **2**, 277.
8. Tomaschewski, J. and Ruger, W. (1987). *Nucleic Acids Res.*, **15**, 3632.
9. Giraud-Panis, E. M-J., Duckett, D. R., and Lilley, D. M. J. (1995). *J. Mol. Biol.*, **252**, 596.
10. Kemper, B., Garabett, M., and Courage, U. (1981). *Eur. J. Biochem.*, **115**, 133.
11. Picksley, S. M., Parsons, C. A., Kemper, B., and West, S. C. (1990). *J. Mol. Biol.*, **212**, 723.

12. Kemper, B., Pottmeyer, S., Solaro, P., and Kosak, H. (1990). *Structure and methods*, Vol. 1, *Human genome initiative and DNA recombination* (Sarma, R. H. and Sarma, M. H., eds), p. 215. Adenine Press, Schenectady, N.Y.

13. Kemper, B., Jensch, F., v.Depka-Prondzynski, M., Fritz, J-J., Borgmeyer, U., and Mizuuchi, K. (1984). *Cold Spring Harbor Symp. Quant. Biol.*, **49**, 815.

14. Jensch, F. and Kemper, B. (1986). *EMBO J.*, **5**, 181.

15. Pottmeyer, S. and Kemper, B. (1992). *J. Mol. Biol.*, **223**, 607.

16. Kleff, S. and Kemper, B. (1988). *EMBO J.*, **7**, 1527.

17. Pottmeyer, S. (1989). *Sequenz-und Strukturspezifität der Endonuclease VII des Bakteriophagen T4. In vitro Untersuchungen an synthetischen DNA Strukturen.* Dissertation an der Universität zu Köln.

18. Bertrand-Burggraf, E., Kemper, B., and Fuchs, R. P. P. (1994). *Mutat. Res. DNA Repair*, **314**, 287.

19. Murchie, A. I. H. and Lilley, D. M. J. (1993). *J. Mol. Biol.*, **223**, 77.

20. Bhattacharyya, A., Murchie, A. I. H., von Kitzing, E., Diekmann, S., Kemper, B., and Lilley, D. M. J. (1991). *J. Mol. Biol.*, **221**, 1191.

21. Birkenkamp, K. and Kemper, B. (1995). *DNA Res.*, **2**, 9.

22. Youil, R., Kemper, B., and Cotton, R. G. H. (1996). *Genomics*, **32**, 431.

23. Youil, R., Kemper, B., and Cotton, R. G. H. (1995). *Proc. Natl. Acad. Sci. USA*, **92**, 87.

24. Guinta, C., Youil, R., Venter, D., Chow, C. W., Somers, G., Lefferty, A., *et al.* (1996). *Diagn. Mol. Pathol.*, **5**, 265.

25. Taylor, G. R. (ed.) (1997). In *Laboratory methods for the detections of mutations and polymorphisms in DNA.* CRC Press, in press.

26. Giunta, C., Youil, R., Knight, M. A., Venter, D., Chow, C. W., Somers, G., *et al.* (1995). *Am. Soc. Hum. Genet.* (Abstract).

27. Mashal, R. D., Koontz, J., and Sklar, J. (1995). *Nature Genet.*, **9**, 177.

28. Babon, J. J., Youil, R., and Cotton, R. G. H. (1995). *Nucleic Acids Res.*, **23**, 5082.

7

Detection of mutations by hybridization with sequence-specific oligonucleotide probes

RANDALL K. SAIKI and HENRY A. ERLICH

1. Introduction

Genetic analysis by hybridization with sequence-specific oligonucleotide probes (SSOP) is one of the oldest methods commonly used to examine nucleic acids for the presence of known sequence variants (1, 2). The procedure takes advantage of the fact that under the appropriate reaction conditions, a short oligonucleotide probe will hybridize to its target only when it is perfectly matched; a single base pair mismatch is often sufficiently destabilizing to prevent a stable probe:target duplex from forming. By testing a specimen with a pair of probes—one complementary to the normal sequence and one to the mutant—the genotype at that position can be determined. Using a panel of probes, an entire series of mutations or polymorphisms can be surveyed.

1.1 The two formats

Today, sequence-specific probe analyses are almost always done in conjunction with a nucleic acid amplification method, typically PCR. The vast enrichment of the DNA sequence of interest that is produced by PCR greatly simplifies subsequent testing by oligonucleotide probes and allows simple detection strategies to be employed. In the classic implementation of SSOP analysis, the nucleic acid to be examined is applied to a solid substrate, usually small strips or sheets of nylon filter membrane, and hybridized with a labelled oligonucleotide in solution. This is known as the immobilized specimen format. Multiple DNA samples can be applied to the membrane, most often in an array using a 96-well, microtitre-style, vacuum spotting manifold (i.e. 'dot blots') (3). If necessary, several replicate filters are prepared, each to be probed with a different oligonucleotide. It is the favoured approach if many specimens are to be tested for the presence of a few sequences. Since each probe is used in separate hybridization reactions, the conditions can be adjusted to suit the individual annealing requirements of the SSOP. However, since it is the DNA sample that is attached to the substrate, new filters must be prepared each time a new specimen is to be tested.

In the reciprocal situation of a moderate number of DNA specimens to be tested for many sequences, the preferred filter organization is one which reverses probe and target: attaching the probes onto the nylon in an array and hybridizing with labelled PCR product in solution. This reverse hybridization approach is called the immobilized probe format. It allows a sample to be analysed by a large number of probes in a single hybridization reaction (4). The oligonucleotides on these filters essentially serve as sequence-specific capture probes, binding and removing only perfectly matched amplicon from solution. As before, multiple membranes can be prepared to be tested with different amplified DNA samples. However, this format is more convenient and provides a simplified typing system. Once a panel of probes is selected, a supply of strips can be made in advance that are individually put to use as amplified specimens become available (which makes it a practical commercial format). In addition, the genotype is more easily determined with these filters since all the probe reactivity data for a particular specimen is contained on one filter. This eliminates a common source of error caused by gathering typing information for a DNA sample spread over multiple filters, as is necessary with the immobilized probe membranes.

1.2 General considerations

SSOP analysis is similar to other types of filter-based hybridization, but re-quires special attention to the hybridization conditions. For the short oligo-nucleotides used in these assays, the difference in melting temperature (T_m) between a perfectly matched sequence and one with a single mismatch is very small, often a matter of only 2–3 °C. Consequently, the salt concentration and temperature (i.e. stringency) must be carefully monitored. Precisely con-trolled and reproducible hybridization stringency is the key to success in this type of analysis. It should be noted, however, that using amplified DNA in these assays greatly improves overall signal intensities, relaxing the require-ments somewhat and introducing a degree of resiliency to the hybridization conditions.

Because of its reliance on perfectly complementary probes, SSOP analysis is best suited for the detection of previously characterized mutants. New sequence variants are more efficiently revealed by one of the scanning tech-niques described in other chapters. A likely exception, however, may be the new 'DNA chip' technology which uses a micro-array of thousands of oligo-nucleotide probes on a small wafer of glass. These chips can exhaustively interrogate a nucleic acid fragment for a host of sequence permutations and generate several hundred bases of sequence data by computer analysis of the resulting hybridization patterns (5).

2. Immobilized specimen filters

Performing a SSOP assay by immobilizing the specimen is the easier of the two formats to set-up initially. It simply involves spotting denatured, PCR amplified specimens onto the nylon filter and hybridizing with one or more labelled oligonucleotide probes. Unless the probe sequences are extensively optimized, the hybridization conditions for each oligonucleotide are likely to require individual adjustment to achieve the best results.

2.1 Designing solution-based sequence-specific oligonucleotide probes

Designing a SSOP and determining the optimal hybridization stringency is a straightforward, but largely empirical, process. Newly made probes must be tested on control samples in order to determine the best hybridization conditions. Under ideal circumstances, a probe should be hybridized to the filter at 2–5 °C below its T_m to obtain good single base mismatch discrimination and adequate signal strength. Interactive computer programs, such as *Oligo 5.0* (National Biosciences), that calculate T_ms (preferably by the nearest-neighbour method) are very helpful during the initial stages of probe design. These programs permit quick evaluation of several candidate probes until one with the appropriate predicted T_m is found. A temperature of 52–55 °C in 1 M monovalent cation is a good starting point for the hybridization conditions described in *Protocol 1*, bearing in mind that the true T_m may be higher or lower by several degrees. The following guidelines should also be considered during probe selection.

2.1.1 Probe length

Probes are usually 15 to 25 nucleotides in length. Base composition and the selected hybridization conditions will determine the optimal length. Oligonucleotides from the lower end of this range often provide better discrimination because a base pair mismatch will have a relatively greater destabilizing effect. Even shorter probes can be used successfully, but they often require hybridization at temperatures below ambient and overall convenience becomes an issue.

2.1.2 Use probe pairs

Whenever possible, probes should be designed and used in pairs—one complementary to the normal sequence and one to the variant. This will assist in establishing the optimal hybridization conditions and provide information about heterozygosity where applicable.

2.1.3 Mismatch position

Mismatches should be placed towards the centre of the oligonucleotide, at least three bases from either end. This maximizes the destabilizing effect of the base pair mismatch on the probe. The type of mismatch is also important.

2.1.4 Avoid G:T (or T:G) mismatches

A G:T pairing is almost neutral and has very little disruptive effect (6). However, if the probe is switched to the other strand, a more potent C:A mismatch is created. The problem of G:T mismatches arises whenever there is a A→G, G→A, T→C, or C→T base substitution. An exercise with paper and pencil will show that whenever strand-switching is employed to avoid a G:T mismatch, a probe pair will have to be complementary to opposite strands (i.e. one 'plus' strand and one 'minus' strand probe) to avoid the reciprocal T:G mismatch.

2.1.5 Be cautious of G:A (or A:G) mismatches

The G:A mismatch is more disruptive than G:T and usually able to prevent probe annealing, but there may be situations (e.g. high per cent GC flanking the mutation site) where the strand-switching strategy is necessary. All other mismatches are strongly destabilizing.

2.1.6 Probe label

Although radioactively labelled probes can certainly be used in these assays, one of the advantages of PCR amplification of the specimen is that simpler, non-radioactive alternatives become possible. Biotinylated oligonucleotides work particularly well as there are numerous avidin-conjugated signal generating systems (e.g. streptavidin–horseradish peroxidase or streptavidin–alkaline phosphatase) and the biotinylated phosphoramidite precursors are readily available. Direct enzyme–probe conjugates (e.g. horseradish peroxidase–oligonucleotide) are somewhat easier to use and resolve the occasional problem of filter background due to non-specific avidin binding, but they can be more difficult to obtain (7).

2.2 Preparation of immobilized specimen filters

Amplified samples are spotted onto the nylon filter as an array of dots using a vacuum manifold. The amplification of specimens is performed by standard procedures. Strategies for selection of primers and optimization of PCR conditions has been described in several publications (8, 9). Only a small amount of amplicon is applied to each dot (5 μl), so prior removal of nucleotide triphosphates and polymerase is unnecessary.

Protocol 1. Preparation of immobilized specimen filters

Equipment and reagents

- Dot blot vacuum manifold (Bio-Rad or Life Technologies)
- Metered UV light box (Stratagene)
- Nylon filter membrane (Pall Biodyne B, 45 μm)
- TE buffer: 10 mM Tris, 0.1 mM EDTA pH 8

- Spotting dye: 0.02% Orange II in TE buffer (Sigma)
- Denaturation solution: 0.4 N NaOH, 20 mM EDTA
- PCR amplified specimen

Method

1. Wet a 9 cm × 13 cm piece of nylon membrane in water. Place in man-ifold following the manufacturer's directions.[a]

2. For each spot, mix 5 µl amplicon, 1 µl spotting dye, and 44 µl de-naturation solution.

3. With vacuum off, dispense 100 µl H_2O into each well to which sample is to be applied.

4. Check to see there are no air bubbles trapped at the bottom of the wells.

5. Dispense denatured samples into wells, mixing well with water.[b]

6. Apply gentle vacuum and draw sample completely through mem-brane.

7. Release vacuum. Dispense 100 µl H_2O into each well. Reapply vac-uum.

8. Remove filter from manifold and irradiate damp membrane with 254 nm UV light to 120 mJ/cm².[c]

9. Rinse thoroughly in water to remove NaOH.

10. Trim excess membrane. Orange dye shows where DNA was spotted.

11. Air dry if not used immediately. Store protected from humidity and bright light.

[a] Exact size of membrane will depend on particular brand of manifold. Size given is appropriate for Bio-Rad apparatus.
[b] Thorough mixing helps dislodge trapped air bubbles.
[c] If metered light source not available, use standard UV light box (type used to photograph ethidium bromide gels). Place filter face down on plastic wrap and expose for 15–45 sec, depending on UV intensity.

A well in which solution flows very slowly or not at all indicates a trapped air bubble at the bottom of the well. Irregular, crescent, or ring-shaped orange spots are also caused by air bubbles.

2.3 Immobilized specimen filter hybridization

The performance of new oligonucleotide probes should be evaluated by hybridization with control DNAs. If non-specific hybridization is observed, stringency should be increased (i.e. lower salt concentration, raise tempera-ture, or increase wash time). Conversely, weak positive signals would suggest decreasing stringency (i.e. raise salt, lower temperature, or decrease wash time). Rather than changing temperature, it is often more convenient to alter the salt concentration. The effect of monovalent cation concentration (e.g. Na^+) on T_m can be estimated from the salt term derived by Wetmur to deter-

Table 1. Effect of salt concentration on T_m

Na$^+$ (molar)	SSPE-equivalent	$\Delta T_m{}^a$
1.2	6 $\times$	0.7
1.0	5 $\times$	0.0
0.8	4 $\times$	−1.0
0.6	3 $\times$	−2.4
0.4	2 $\times$	−4.6
0.2	1 $\times$	−8.7

^{a} ΔT_m normalized for 1.0 M salt (5 $\times$ SSPE).

mine melting temperatures for oligonucleotides (10) and is given in *Table 1* for several cases.

$$T_m \sim 16.6 \log \frac{[\text{Na}^+]}{1.0 + 0.7\,[\text{Na}^+]}$$

Protocol 2 describes a common set of hybridization conditions. Your own probes may require different salt or temperature.

Protocol 2. Hybridization of probes to immobilized specimen filters

Equipment and reagents

- Shaking water-bath set at 50°C (Bellco)
- Accurate thermometer
- Plastic hybridization tray (Perkin Elmer)^a
- 20 $\times$ SSPE: 3.6 M NaCl, 0.2 M NaH$_2$PO$_4$, 0.02 M Na$_2$EDTA, adjusted to pH 7.4 with 10 M NaOH
- Hybridization buffer: 5 $\times$ SSPE, 0.5% SDS
- Wash buffer: 2.5 $\times$ SSPE, 0.1% SDS
- Hydrogen peroxide: 3% in water
- Enzyme conjugate: streptavidin–horseradish peroxidase (Perkin Elmer)
- Development buffer: 100 mM trisodium citrate, adjusted to pH 5 with citric acid
- Chromogen: 2 mg/ml 3,3′,5,5′-tetramethylbenzidine in 100% ethanol (Perkin Elmer or Fluka)
- Biotinylated oligonucleotide probe

Method

1. Check temperature of shaking water-bath with thermometer.
2. Warm hybridization and wash buffers slightly in 37°C water-bath or incubator. Check to see detergent is in solution.
3. Place filter in tray and add 25 ml hybridization buffer. Add 10 pmol (~ 60 ng 20-mer) of biotinylated probe.
4. Set tray in shaking water-bath and agitate gently for 20 min.
5. Remove tray and aspirate hybridization solution.
6. Rinse filter in 25 ml wash buffer and aspirate.

118

7. Add another 25 ml wash buffer and return tray to shaking water-bath for 10 min.[b]

8. Remove tray and aspirate wash buffer.

9. Mix 25 ml wash buffer and 50 μl enzyme conjugate. Add to tray. Place tray on platform shaker and rotate gently at room temperature for 20 min.

10. Aspirate wash buffer. Add 25 ml wash buffer and return to platform shaker for 10 min.

11. Aspirate wash buffer. Add 25 ml development buffer and return to platform shaker for 10 min.

12. Repeat step 11 once. Aspirate development buffer.

13. Mix 1 ml chromogen with 20 ml development buffer and 20 μl hydrogen peroxide. Add to tray. Develop colour for 10–30 min at room temperature with gentle shaking.

14. Rinse thoroughly in water to stop colour development of the blue dots.

15. Photograph for permanent record.

[a]Two sizes of trays are available from the manufacturer. One has capacity for a 8 × 12 array of dots (microtitre plate), the other a single row of up to 12 dots. The protocol assumes the larger tray. If the smaller tray is used, reduce volumes fivefold.
[b]The stringent wash is the most important step in the assay. Both time and temperature must be carefully controlled to obtain reliable results.

If desired, multiple hybridization conditions can be tested simultaneously by hybridizing several replicate filters at the same temperature under slightly different salt concentrations. For example, the series 3 ×/1.5 ×, 4 ×/2 ×, 5 ×/2.5 ×, and 6 ×/3 × represent the SSPE concentrations for hybridization and wash buffers, respectively, spanning approximately 4 °C T_m range. It is not uncommon for the optimal hybridization conditions for each member of a normal/mutant pair of oligonucleotide probes to be slightly different. If multiple hybridization conditions are unacceptable, it will be necessary to redesign the out-of-range probes, typically by adding or subtracting one or more bases.

Various streptavidin–enzyme conjugates are available from several vendors, but may exhibit considerable batch-to-batch variability in filter-based applications. The streptavidin–horseradish peroxidase from Perkin Elmer is specifically prepared for use in these types of assays.

2.4 Reusing immobilized specimen filters

Depending on the signal-generating method used, the strips can often be reused by removing the bound probe and, if necessary, the signal itself. For the tetramethylbenzidine (TMB) system used in *Protocol 2*, the filters are first decolorized then heated to remove the probe. For filters destined to be reused, it is best not let them dry out or expose them to UV light (treatments that tend to fix DNA to nylon), before the probes are removed. In some cases where the blue TMB signal is exceptionally dark, it may not be possible to completely decolorize or the dot may turn yellow. Strips in this condition probably cannot be reused.

Protocol 3. Removal of probe from the filter

Equipment and reagents

- Microwave oven or hot plate
- Glass crystallizing dish and watch glass
- TE buffer (*Protocol 1*)

- Decolorization solution: 1% sodium sulfite in water
- Stripping solution: 0.1% SDS in TE buffer

Method

1. Place filter in crystallizing dish and add enough decolorization solution to cover.

2. Incubate at room temperature with moderate agitation until decolorization is complete (10–60 min).[a]

3. Rinse filter in water.

4. Add stripping solution and cover dish with watch glass.

5. Heat to gentle boil in microwave or on hot plate for 5–10 min.

6. Repeat steps 3–5.

7. Rinse in water and air dry for storage.

[a] If colour remains, increase concentration of sodium sulfite to 5% and incubate for longer period (up to overnight).

---

3. Immobilized probe filters

The immobilized probe format is very convenient in situations where the specimen is to be tested with multiple probes. Since the probes are attached to the substrate in a defined array, the DNA sample can be screened with all SSOPs in a single hybridization. However, unlike the immobilized specimen filter where small adjustments can be made for the optimal annealing of each probe, all the probes on an immobilized probe filter must hybridize specifically under identical conditions. This often requires several revisions of a particular oligonucleotide probe in an effort to identify one with the appropriate hybridization characteristics. This aspect can prove to be quite challenging. Nevertheless, once a panel of probes is developed, the application of the immobilized probe format is very simple and reliable. The main advantage is in the ability to prepare a large quantity of filters ahead of time and use them

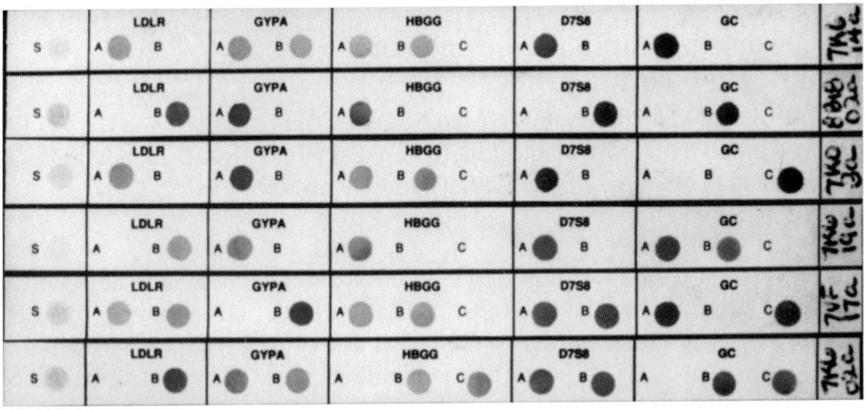

Figure 1. A commercial immobilized probe filter used to genotype six individuals for forensic purposes. The AmpliType PM strips (Perkin Elmer) simultaneously examines five polymorphic genetic loci—LDLR, GYPA, HBGG, D7S8, and GC. Genotype is determined by simply inspecting the pattern of dots. For example, the genotype of the specimen hybridized to the top filter would be, LDLR: A/A, GYPA: A/B, HBGG: A/B, D7S8: A/A, GC: A/A. These strips are made with synthetically tailed probes on nylon membrane with silk-screened lettering. The primer set also amplifies the second exon of the HLA-DQA1 locus.

as they are needed. Even for moderately high volume tests, like HLA typing, where the number of specimens will readily exceed the number of probes, it may still be preferable to use immobilized probe strips because of ease of use (11).

Probes can be attached to the nylon filter by several means. All of them involve the addition of a discrete attachment moiety. Although a method for direct covalent attachment by means of 5' reactive amino groups has been described (12), the addition of a 3' polydeoxythymidine tail with terminal transferase is a straightforward treatment that can be applied to any candidate probe with an available 3' hydroxyl group (4). UV irradiation activates thymine bases in DNA which then can react with the primary amines of the nylon membrane (13). Consequently, a long poly(dT) tail of several hundred bases serves as both a preferred attachment point and as a spacer to separate the probe from the filter surface. Except for probe tailing, the other procedures are very similar to those used for the immobilized specimen filters (*Protocols 1–3*). An example of an immobilized probe filter used in forensic applications is shown in *Figure 1*.

Introducing a label into the PCR product can be accomplished with either labelled primers or nucleotide triphosphates. Biotinylated primers are readily synthesized and convenient to use. Biotinylated triphosphates may require a little more experimentation to determine optimum substitution levels, but can result in increased sensitivity by incorporating multiple biotins per DNA strand.

3.1 Designing immobilized sequence-specific oligonucleotide probes

The process of designing immobilized SSOPs is very similar to that used for solution-based probes. In addition to the guidelines listed in Section 2.1, immobilized probes are affected by two more considerations.

3.1.1 Adjacent sequence

For probes to be tailed with poly(dT), the sequence immediately adjacent to 3′ end on the probe at the site of hybridization in the DNA sample must be examined for complementarity to the tail. If one or more successive adenines are present, they will hybridize to the tail and increase the effective length of the probe, potentially creating a probe with a higher T_m than intended. In these situations, the problem can be overcome by incorporating a non-complementary base, such as dC, into the oligo during synthesis. The result will be a compound tail of one or more cytosines followed by the poly(dT) tract.

Alternatively, 3′ thymidines might prove to be necessary in a particular SSOP. Although it is certainly possible to rely upon the homopolymer tail to provide the missing terminal thymidines, it is more prudent to make these bases part of the synthesized probe so the intention of the designer is clear.

3.1.2 Amplicon secondary structure

Since the PCR products in this format are not bound to the filter, the single-stranded amplicon in solution is free to form secondary structures. On occasion, a particularly stable hairpin may block the binding site and prevent the probe from hybridizing. Many times, simply increasing the length of the probe will allow it to successfully compete with intramolecular structures. Even though the predicted T_m for a longer probe may make it appear too high for the intended reaction conditions, the oligonucleotide is in competition with another part of the amplicon which lowers the observed T_m. The effect of secondary structure often results in a panel of probes with a wide range of calculated T_ms, although in practice they all hybridize properly under one set of conditions.

Another approach is to examine the amplicon sequence with a DNA folding program to determine where the other half of the hairpin originates and then relocating the primers to avoid it. This strategy is less successful, probably because in genes with high levels of secondary structure, like the human β-globin gene, merely moving the primers allows other hairpins to form.

3.2 The probe tailing reaction

The addition of a homopolymer tail to the oligonucleotides is conveniently accomplished with terminal transferase. The length of the tail is controlled by adjusting the molar ratio of probe and nucleotide triphosphate in the reaction.

Enzymatically synthesized tails are usually about 400 bases long. The longer tails allow better hybridization efficiency, probably because of a spacer effect that separates the probe from the membrane surface (4).

An alternative, non-enzymatic way to add a polynucleotide tail onto the end of a probe is to create it during synthesis. In this case, the tails are only 50 to 200 bases and appended to the 5' end of the oligonucleotide. To improve the yield, the capping reaction can be disabled on the synthesizer during the addition of the homopolymer since a few missing thymine bases will not affect the performance of the tail and there is a significant increase in synthesis efficiency. (DNA is synthesized in the 3' to 5' direction, so the probe will be made first followed by the tail.)

Protocol 4. Tailing probes with terminal transferase

Equipment and reagents

- Terminal deoxyribonucleotidyl transferase (TdT): ~ 20 U/µl (Pharmacia)
- 10 × TdT buffer: 1000 mM potassium cacodylate, 10 mM CoCl₂, 2 mM dithiothreitol, 250 mM Tris pH 7.6[a]
- 8 mM dTTP pH 7 (Pharmacia)
- TE buffer (*Protocol 1*)
- Oligonucleotide probe: 10 µM in TE buffer
- Stop buffer: 10 mM EDTA pH 8

Method

1. Combine 10 µl 10 × TdT buffer, 10 µl dTTP (80 nmol), 20 µl oligonucleotide probe (0.2 nmol), and 60 µl dH₂O.[b]

2. Add 1 µl (~ 20 U) terminal transferase. Incubate overnight at 37 °C.

3. Add 100 µl stop buffer. Store at –20 °C.

[a] See ref. 14 for details of preparation.
[b] If desired, tail length can be adjusted by changing the molar ratio of dTTP to oligonucleotide.

3.3 Preparation of immobilized probe filters

Tailed probes are applied to the nylon filter essentially as described in Section 2.2. However, less UV irradiation is required to fix the probes to the membrane because of the reactive polythymidine tract.

Protocol 5. Preparation of immobilized probe filters

Equipment and reagents

- Dot blot vacuum manifold (Bio-Rad or Life Technologies)
- Metered UV light box (Stratagene)
- Nylon filter membrane (Pall Biodyne B, 45 µm)
- TE buffer (*Protocol 1*)
- Spotting dye (*Protocol 1*)
- Poly(dT) tailed oligonucleotide probe (1 µM, from *Protocol 4*)

Protocol 5. *Continued*

Method

1. Wet a 9 cm × 13 cm piece of nylon membrane in water. Place in manifold following the manufacturer's directions.[a]

2. For each spot, mix 5 μl tailed probe (5 pmol), 1 μl spotting dye, and 44 μl TE buffer.

3. With vacuum off, dispense 100 μl TE buffer into each well to which sample is to be applied.

4. Check to see there are no air bubbles trapped at the bottom of the wells.

5. Dispense diluted probe into wells, mixing well with TE buffer.[b]

6. Apply gentle vacuum and draw sample completely through membrane.

7. Release vacuum. Dispense 100 μl TE buffer into each well. Reapply vacuum.

8. Remove filter from manifold and irradiate damp membrane with 254 nm UV light to 60 mJ/cm^2.[c]

9. Rinse thoroughly in water.

10. Trim excess membrane. Orange spotting dye shows where probe was spotted.

11. Air dry if not used immediately. Store protected from humidity and bright light.

---

[a] Exact size of membrane will depend on particular brand of manifold. Size given is appropriate for Bio-Rad apparatus.
[b] Thorough mixing helps dislodge trapped air bubbles.
[c] If metered light source not available, use standard UV light box (type used to photograph ethidium bromide gels). Place filter face down on plastic wrap and expose for 10–30 sec, depending on UV intensity.

3.4 Immobilized probe filter hybridization

Once again, the procedure for hybridizing immobilized probes filters is similar to that for immobilized specimens.

Protocol 6.   Hybridization of amplicon to immobilized probe filters

Equipment and reagents

- Shaking water-bath set at 50°C (Bellco)
- Accurate thermometer
- Plastic hybridization tray (Perkin Elmer)[a]
- Denaturation solution (*Protocol 1*)
- Hybridization buffer (*Protocol 2*)
- Wash buffer (*Protocol 2*)

- Enzyme conjugate (*Protocol 2*)
- Development buffer (*Protocol 2*)
- Chromogen (*Protocol 2*)
- Hydrogen peroxide (*Protocol 2*)
- Biotinylated, PCR amplified specimen

Method

1. Check temperature of shaking water-bath with thermometer.

2. Warm hybridization and wash buffers slightly in 37 °C water-bath or incubator. Check to see detergent is in solution.

3. For each filter, mix equal volumes amplicon and denaturation solution (10–35 μl each).

4. Place filter in tray and add 5 ml hybridization buffer. Add denatured amplicon.

5. Set tray in shaking water-bath and agitate gently for 20 min.

6. Remove tray and aspirate hybridization solution.

7. Rinse filter in 5 ml wash buffer and aspirate.

8. Add another 5 ml wash buffer and return tray to shaking water-bath for 10 min.[b]

9. Remove tray and aspirate wash buffer.

10. Mix 5 ml wash buffer and 10 μl enzyme conjugate. Add to tray. Place tray on platform shaker and rotate gently at room temperature for 20 min.

11. Aspirate wash buffer. Add 5 ml wash buffer and return to platform shaker for 10 min.

12. Aspirate wash buffer. Add 5 ml development buffer and return to platform shaker for 10 min.

13. Repeat step 11 once. Aspirate development buffer.

14. Mix 0.1 ml chromogen with 2 ml development buffer and 2 μl hydrogen peroxide. Add to tray. Develop colour for 10–30 min at room temperature with gentle shaking.

15. Rinse thoroughly in water to stop colour development of the blue dots.

16. Photograph for permanent record.

[a]Two sizes of trays are available from manufacturer. One has capacity for a 8 × 12 array of dots (microtitre plate), the other a single row of up to 12 dots. The protocol assumes the smaller tray. If the larger tray is used, increase volumes fivefold.
[b]The stringent wash is the most important step in the assay. Both time and temperature must be carefully controlled to obtain reliable results.

3.5 Reusing immobilized probe filters

Reusing strips with immobilized probes is possible, but is unreliable. The problem appears to be the susceptibility of the bound PCR product to become attached to the filter, where the single-stranded portion of the amplicon not duplexed with the probe comes into contact with the nylon substrate. As mentioned in Section 2.4, filters with bound amplicon absolutely should not be allowed to dry or be exposed to UV light. The stripping procedure (*Protocol 3*) should be performed as soon as possible to limit the possible interaction with the nylon. Processed filters can be tested by colour development (*Protocol 6*, steps 13–15) to confirm that the biotin labelled PCR product was, in fact, removed.

4. DNA chips

The so-called DNA chips represent a new technology with considerable potential. These glass or silicon chips contain micro-arrays of oligonucleotides, essentially a miniaturized version of an immobilized probe membrane. There are several companies at work in this field, each with different strategies for manufacturing chips. One of these organizations is Affymetrix. The oligonucleotides in the Affymetrix array are synthesized directly on glass chips using a photoreactive phosphoramidite chemistry and photolithographic masking technology adapted from the semiconductor industry (15). Individual probes occupy array elements of only 50–100 μm square. A 1.28×1.28 cm chip with array elements of 100 μm can hold over 16 000 different probes or 65 000 probes with 50 μm elements. The chips are hybridized to fluorescently labelled PCR product then scanned by confocal laser microscopy. A computer interprets the resulting complex pattern of probe reactivity to determine the genotype.

Because of the vast capacity of these chips, it is possible to create an overlapping series of SSOPs that test for all possible base substitutions at each position over an extended region of DNA. Single base deletions can also be tested for with a fifth oligonucleotide. Using this approach, the entire human mitochondrial genome of 16.6 kb has been sequenced on one chip (16).

While the Affymetrix technology is very good at identifying single base substitutions, it is not as proficient at handling base insertions or deletions. The arrays also have difficulties in situations where multiple sequence changes occur in close proximity. However, the probes will indicate the precise location of these anomalies by failing to hybridize and another method can be used to determine the exact nature of the mutation.

As in a traditional filter-based SSOP analysis, chip oligonucleotides can be designed to detect specific mutations but with substantially increased probe capacity. Instead of designing one optimal probe for each normal and mutant sequence, an entire series of oligonucleotides test each region and it is left to the computer to sort the data out. In practice, the analysis amounts to a minisequencing around the mutation position (17). *Figure 2* is an example of a 0.5×0.5 cm chip that tests for 58 specific cystic fibrosis mutations. Each

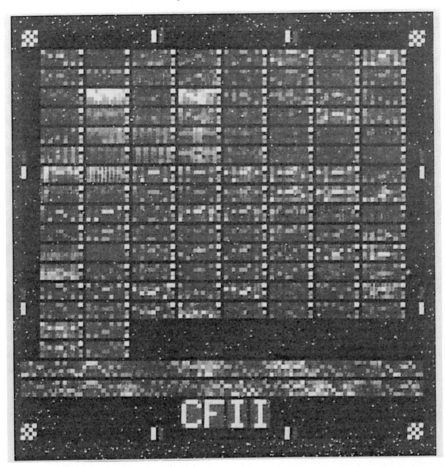

Figure 2. The 58 mutation cystic fibrosis chip made by Affymetrix. Each probe 'element' is 50 μm square. The entire probe array is 5.25 mm square and consists of 4640 allele-specific probes (58 mutations, 20 probes per sequence, both strands, normal and mutant probe sets: 58 × 20 × 2 × 2), not including ~ 500 control probes. In addition, the two horizontal blocks near the bottom of the chip contain 840 overlapping probes (105 positions, four probes per position, both strands: 105 × 4 × 2) that are used to sequence a 105 bp region at exon 11 of the *CF* gene. One of the challenges in designing assays for these chips is developing the high multiplex PCRs, such as the 20-plex reaction used here.

mutation location is probed by a panel of 80 oligonucleotides: 40 for the normal sequence and 40 for the mutant.

5. Applications

The first diagnostic application of the PCR/SSOP approach to genetic typing was detection of haemoglobinopathies, with amplification of the β-globin locus and the direct identification of the β^A allele and the mutant β^S and β^C alleles in the dot blot format (3); this report also included the analysis of HLA polymorphism with the identification of alleles at the HLA-DQA1 locus. This PCR/dot blot test for HLA-DQA1 typing was used in 1986 in the first DNA forensic case in the US (Pennsylvania versus Pestinikis) (18) and, in the immobilized probe format (4), the HLA-DQA1 test became the first commercial PCR-based kit, the Amplitype™ DQα PCR Amplification & Typing Kit. Since its appearance in 1990, this reverse dot blot test has been used in thousands of forensic cases around the world. Results obtained with the subsequent immobilized probe test for forensic genetic typing, a multiplex amplification system for six loci, the Amplitype PM PCR Amplification & Typing Kit are shown in *Figure 1*.

In addition to the HLA-DQA1 test, the immobilized probe format has also been used extensively in HLA tissue typing for transplantation for the other HLA loci. The HLA class I and class II loci are extremely polymorphic with several loci (i.e. HLA-DRB1 and HLA-B) having well over 100 alleles. For the class II loci, the polymorphism is localized to the second exon and, for the class I loci, to the second and third exons. This allelic diversity requires many probes for genetic typing. In order to accommodate a larger number of probes on the membranes, for more recently developed tests, the probes have been deposited as lines rather than as dots. The commercial test for the HLA-DRB1, DRB3, DRB4, and DRB5 loci, the Amplicor DRB test, is a 'line blot' with 29 probes (11). The test for the HLA-A locus is a line blot with 52 probes (19) while the prototype for HLA-B typing has over 80 probes. Because of the patchwork pattern of polymorphism for both the HLA class I and class II loci, new (previously unreported) sequence variants can be identified as a new combination of polymorphic sequence motifs and, thus, detected as a novel pattern of probe reactivity. To facilitate HLA typing using the immobilized probe format, instruments (SLT Profiblot) for automating the hybridization, washing, and chromogenic detection of the strips have been developed. In addition, software has been developed to allow the scanning of the probe reactivity patterns with a conventional flat-bed scanner and the interpretation of this pattern as an HLA genotype.

Other genetic typing tests, such as the detection of 16 CFTR mutations, confirmatory tests for newborn haemoglobinopathy screening in California, and forensic tests for mitochondrial DNA polymorphism have been developed in the immobilized probe format. Genetic typing for pathogens, such as the identification of human papilloma virus (HPV) types has also been recently developed as a 29 probe line blot.

6. Summary/perspectives

The method of PCR amplification followed by analysis with SSO probes has proved to be a powerful and general approach to genetic typing for the identification of previously characterized sequence variants. The immobilized probe format provides a very convenient, simple, and robust method of mutation/polymorphism detection for clinical diagnosis and one that can be automated for high throughput analysis. However, the nylon membrane probe arrays discussed above are limited in the number of probes that can be accommodated on a single strip and are generally not capable of identifying unknown mutations and sequence variants. The potential of DNA chips that can contain thousands of probes in a micro-array promises to extend PCR/SSO analysis to a wide variey of research and clinical applications.

References

1. Wallace, R. B., Shaffer, J., Murphy, R. F., Bonner, J., Hirose, T., and Itakura, K. (1979). *Nucleic Acids Res.*, **6**, 3543.
2. Conner, B. J., Reyes, A. A., Morin, C., Itakura, K., Teplitz, R. L., and Wallace, R. B. (1983). *Proc. Natl. Acad. Sci. USA*, **80**, 278.
3. Saiki, R. K., Bugawan, T. L., Horn, G. T., Mullis, K. B., and Erlich, H. A. (1986). *Nature*, **324**, 163.
4. Saiki, R. K., Walsh, P. S., Levenson, C. H., and Erlich, H. A. (1989). *Proc. Natl. Acad. Sci. USA*, **86**, 6230.
5. Kozal, M. J., Shah, N., Shen, N., Yang, R., Fucini, R., Merigan, T. C., *et al.* (1996). *Nature Med.*, **2**, 753.
6. Ikuta, S., Takagi, K., Wallace, R. B., and Itakura, K. (1987). *Nucleic Acids Res.*, **15**, 797.
7. Saiki, R. K., Chang, C. A., Levenson, C. H., Warren, T. C., Boehm, C. D., Kazazian, H. H. Jr., *et al.* (1988). *N. Engl. J. Med.*, **319**, 537.
8. Erlich, H. A. (ed.) (1989). *PCR technology: principles and applications for DNA amplification.* Stockton Press, NY.
9. Innis, M. A., Gelfand, D. H., and Sninsky, J. J. (1995). *PCR strategies.* Academic Press, London.
10. Wetmur, J. G. (1991). *Crit. Rev. Biochem. Mol. Biol.*, **26**, 227.
11. Begovich, A. and Erlich, H. (1996). *J. Am. Med. Assoc.*, **273**, 586.
12. Zhang, Z., Coyne, M. Y., Will, S. G., Levenson, C. H., and Kawasaki, E. S. (1991). *Nucleic Acids Res.*, **19**, 3929.
13. Church, G. M. and Gilbert, W. (1984). *Proc. Natl. Acad. Sci. USA*, **81**, 1991.
14. Roychoudhury, R. and Wu, R. (1980). In *Methods in enzymology* (ed. L. Grossman and K. Moldave) Vol. 65, p. 43. Academic Press, London.
15. Pease, A. C., Solas, D., Sullivan, E. J., Cronin, M. T., Holmes, C. P., and Fodor, S. P. A. (1994). *Proc. Natl. Acad. Sci. USA*, **91**, 5022.
16. Chee, M., Yang, R., Hubbell, E., Berno, A., Huang, X. C., Stern, D., *et al.* (199x). *Science*, **274**, 610.
17. Cronin, M. T., Fucini, R. V., Kim, S. M., Masino, R. S., Wespi, R. M., and Miyada, C. G. (1996). *Hum. Mutat.*, **7**, 244.
18. Blake, E., Mihelovich, J., Higuchi, R., Walsh, P. S., and Erlich, H. A. (1992). *J. Forensic Sci.*, **37**, 700.
19. Bugawan, T. L., Apple, R., and Erhlich, H. A. (1994). *Tissue Antigens*, **44**, 137.

8

DNA detection and sequence distinction through oligonucleotide ligation

ULF LANDEGREN, MARTINA SAMIOTAKI,
MAREK KWIATKOWSKI, JÜRI PARIK, MATS NILSSON,
ANETTE HAGBERG, and GISELA BARBANY

1. Introduction

Synthetic oligonucleotides and DNA ligases are combined in a number of methods for gene analysis. Generally, two oligonucleotide probes are designed to hybridize immediately next to each other on a target DNA or RNA strand. Ends of the oligonucleotide probes, brought close by hybridization, can be covalently joined by a DNA ligase, provided that the probes are properly base paired to the target sequence at the junction. This joining act thus reflects the presence of the target sequence. Below, a method will be presented to use oligonucleotide ligation-assisted analysis to distinguish known variants of amplified DNA sequences.

There are three principal features of oligonucleotide ligation assays that render them valuable in many types of genetic analyses:

(a) Sequences are detected with high specificity, due to the requirement that two different target sequences are recognized before ligation of the two probes can take place, and a positive reaction is scored. This specificity is adequate to analyse single copy sequences in complex genomic DNA samples.

(b) Closely similar sequence variants can be efficiently distinguished, because of the substrate requirements of DNA ligases that allows probe:target hybrids with matched and mismatched ends to be distinguished.

(c) The natural phosphodiester bond, formed between oligonucleotide probe ends that have been joined by ligation, provides a stable link and allows the joined probe pairs to serve as templates for another ligation reaction, or for replication. This last property can be exploited in different ways to distinguish a small number of ligated probes from an excess of unreacted probes.

2. Background

2.1 A brief history of ligation assays

Besmer and co-workers first applied the enzymatic joining of two oligonucleotides hybridized to a tRNA as a means of characterizing the target molecule (1). In 1988 two papers appeared, demonstrating that the act of ligation serves to distinguish sequence variants, as mismatched probes ends fail to be joined, or are joined considerably less efficiently, when mismatched to the target sequence (2, 3). This observation was extended by further characterizing the optimal conditions for target distinction by ligation (4). A ligase chain reaction was established as a method to combine accurate target detection with an amplified signal (5, 6). Several reviews exist that describe the methodology in greater detail (7–9).

2.2 Target sequence distinction by DNA ligases

Ligation assays can take advantage of the lower hybridization stability of probes mismatched to their target sequences in order to distinguish similar target sequence variants (10). It is, however, more convenient to use as the basis for target sequence distinction the reduced ability of oligonucleotide to serve as a substrate for enzymatic joining when the ends to be joined are mismatched to their target (2, 3). Conditions to distinguish point mutation using the *Thermus thermophilus* enzyme have been carefully optimized, providing a differential ligation of matched over single nucleotide mismatched junctions of more than 40 000-fold (11).

2.3 The oligonucleotide ligation assay (OLA)

The joining of pairs of oligonucleotides as a consequence of their hybridization to a target molecule can be conveniently monitored by providing one of the oligonucleotides with a detectable functional group, and the other with a group that can be bound to a solid phase either before or after the ligation reaction (*Figure 1*). This detection technique offers a number of valuable features:

(a) The covalent bond formed by ligation resists superstringent washes, serving to reduce background.

(b) Since labels attached to the oligonucleotides can be placed in a region remote from that recognized by the enzyme, bulky groups may be used in order to ensure sensitive detection. Moreover, standard conditions may be used for the detection of any mismatched target sequence from a correct match, allowing for easy automation of the assay (12).

(c) Finally, by using three different ligation probes, two of them specific for the normal or the mutant variant of a target sequence, respectively, and differentially labelled, alternative sequence variants may be compared in a single reaction, ensuring increased precision of analysis and reduced susceptibility to experimental artefacts (13, 14). A protocol for such an assay is given in Section 3.

132

3. DNA sequence distinction by PCR followed by OLA

3.1 PCR amplification

DNA samples to be analysed for allelic sequence variation in a given position can be amplified by PCR, followed by ligase-mediated gene detection. Amplification reactions are performed directly in microtitre wells or they can be performed in microcentrifuge tubes. A 4 μl amplification reaction is used for each ligation reaction. In order to amplify samples in a total volume of 4 μl, mix equal volumes of genomic DNA and of *Taq* polymerase and primers both in 1 × PCR buffer. In this manner, fluctuations in buffer composition due to pipetting errors are avoided.

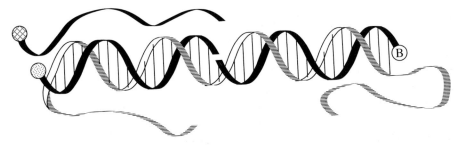

Figure 1. Oligonucleotide ligation analysis of allelic sequence variants in amplified DNA. After target amplification by PCR a set of three oligonucleotides and a ligase are added, and the ligation products are collected on a solid support for detection. The two upstream oligonucleotide probes are differentially labelled and are specific for different target sequence variants. The downstream oligonucleotide is modified with a group (B) allowing it to be bound to a solid phase before or after the ligation reaction.

Protocol 1. Amplification

Equipment and reagents

- Thermal cycler suitable to amplify samples in microtitre plates (MJ Research or Perkin Elmer Cetus)
- Oligonucleotide primers

- 1 × PCR buffer: 50 mM Tris–HCl pH 8.3, 50 mM KCl, 1.5 mM MgCl$_2$, 12.5 μg/ml BSA, 200 μM dNTP
- *Taq* polymerase (Perkin Elmer Cetus)

Method

1. Add to each well of a microtitre plate 2 μl of genomic DNA at 2 ng/μl in 1 × PCR buffer.

2. Next add 2 μl of *Taq* polymerase (0.2 U/μl) and primers (2 μM each), also in 1 × PCR buffer.

3. Overlay with a drop of mineral oil.

4. Subject to 30 temperature cycles, typically 94°C, 55°C, and 72°C.

3.2 Ligation reaction

After amplification the reactions are diluted with water and heated to denature the PCR products. After the temperature has been reduced to 37°C a ligation mix, including three ligation probes and a ligase, is added to individual amplification reactions.

Protocol 2. Ligation

Reagents

- Ligation mix containing per reaction: 1.8 μl of 100 mM Tris–acetate pH 7.5, 100 mM magnesium acetate, 500 mM potassium acetate, 0.8 μl 5 M NaCl, 0.2 μl 100 mM ATP, 0.4 mU of T4 DNA ligase (Pharmacia)
- A set of three labelled oligonucleotides (600 fmol each): one biotinylated and two allele-specific ones, differentially labelled with europium or terbium chelates

Method

1. Amplification reactions are diluted with 6 μl water to a volume of 10 μl.
2. The reactions are heated to 96°C for 5 min to denature the PCR products.
3. The temperature is then rapidly lowered to 37°C.
4. After the solutions have reached a temperature below 50°C, 10μl of a ligation mix is added to individual amplification reactions for a final volume of 20μl.
5. The ligation reactions are incubated for 30 min at room temperature or at 37°C.

3.3 Binding of ligation products to a solid support

In order to isolate ligation products for subsequent washes and detection these are captured on a solid support. Several types of streptavidin-coated solid supports are suitable to use in the assay: streptavidin-coated paramagnetic particles (Dynabeads, Dynal AS) can be used. As an alternative, the ligation reactions may be added to microtitre plates, previously coated with streptavidin. We prefer to handle individual ligation reactions by introducing prongs of a streptavidin-coated manifold support into all 96 wells of a microtitre plate in parallel, thereby avoiding the requirement to pipette individual reactions. While streptavidin-coated paramagnetic particles and microtitre plates are commercially available, manifold supports have to be constructed as described below (see also ref. 15). The capture protocol below works well with either of these supports.

By fixing porous particles to the manifold prongs as described above, the surface available for interaction is increased by almost three orders of magnitude and binding capacities of several tens of picomoles of biotin are routinely obtained (15). It is prudent to test the binding capacity of newly constructed supports. A convenient means of conducting such tests is by adding increasing amounts of free biotin to wells containing a fixed amount of a detectable biotinylated molecule. That amount of biotin that reduces the binding by one-half is taken as a measure of the binding capacity of the supports.

Protocol 3. Construction of manifold supports

Equipment and reagents

- Ultrasonic water-bath
- Manifold supports: configured as a microtitre plate lid with eight rows of 12 ball-and-pin extensions projecting into individual wells of a microtitre plate located underneath are available from Falcon, Oxnard CA
- The particles used for immobilization are Sepharose particles: HiTrap, NHS-activated HP Sepharose, available as a suspension in isopropanol (Pharmacia Biotechnology)

- 1 mM HCl
- 1.0 M NaCl, 0.4 M NaHCO$_3$ pH 8.3, containing 4 mg/ml of streptavidin
- 0.1 M ethanolamine–HCl buffer pH 8.3
- 0.1 M acetate buffer pH 4.0
- 50 mM Tris–HCl pH 7.3 with 0.02% (w/v) sodium azide
- Methanol
- Triethylamine

A. *Coupling of streptavidin to porous agarose particles*

1. 6 ml of sedimented Sepharose particles are washed for less than 1 min with ice-cold 1 mM HCl (3 × 10 ml) on a sintered funnel.

2. The particles are transferred to 5 ml of 1.0 M NaCl, 0.4 M NaHCO$_3$ pH 8.3, containing 20 mg of streptavidin.

3. The suspension is incubated rotating end-over-end for 1 h, filtered, and the excess active sites on the particles are blocked in 0.1 M ethanolamine–HCl buffer pH 8.3 for 15 min.

4. Streptavidin-conjugated Sepharose particles are washed with 0.1 M acetate buffer pH 4.0, followed by distilled water, and used immediately. Alternatively, batches of conjugated particles can be kept at 4°C until use in a storage buffer of 50 mM Tris–HCl pH 7.3 with 0.02% (w/v) sodium azide.

B. *Annealing of agarose particles to plastic supports*

1. The particles are dried on the filters by washing with methanol (3 × 5 ml), followed by triethylamine (3 × 5 ml), and then suspended as a 50% (v/v) slurry in triethylamine.

2. The polystyrene support is washed with ethanol for 20 min in an ultrasonic bath and air dried.

3. The prongs of the support are immersed in the slurry for 2 sec, air dried, and the procedure is repeated once more.

4. The supports are washed under running tap-water.

5. The supports are kept in storage buffer in a humidified container at 4°C until they are used. Alternatively, they may be stored dry.

Manifold supports are useful in many assays besides oligonucleotide ligation and serve to avoid manipulations of individual samples with concomitant risks of sample mix-up and contamination. Besides streptavidin other proteins such as antibodies may be conveniently attached in a similar fashion, as well as oligonucleotides, for instance for mRNA isolation.

Protocol 4. Capture on support

Equipment and reagents

- Shaking platform (e.g. Wallac, Finland)
- Streptavidin-coated paramagnetic particles (Dynabeads, Dynal AS) or, alternatively, manifold supports (see *Protocol 3*)
- Solution A: 1 M NaCl, 100 mM Tris–HCl pH 7.5, 0.1% Triton X-100
- Denaturing solution: 0.1 M NaOH, 1 M NaCl, 0.1% Triton X-100

Method

1. After ligation, 2 μl of streptavidin-coated paramagnetic particles are added per reaction. Alternatively one prong of a manifold support is immersed in each microtitre well.
2. The reactions are incubated on a shaking platform at room temperature for at least 15 min.
3. The supports are washed twice with solution A, then once with a denaturing solution, and finally twice more with buffer A. Paramagnetic particles are washed by attracting to one side of the wells using a permanent magnet, while manifold supports are washed by removing from reaction wells and immersing the prongs in the washing solution.

3.4 Detection of ligation products

The support-bound oligonucleotides, labelled with europium or terbium chelates, are detected by time-resolved fluorescence measurement.

Protocol 5. Detection by time-resolved fluorometry

Equipment and reagents

- Delfia Plate Reader Research Fluorometer (Wallac)
- Fluorescence enhancement solution for europium or samarium: 0.1 M acetate–phthalate pH 3.2, 15 mM 2-naphthoyl trifluoroacetone, 50 mM tri-*N*-octylphosphine oxide, and 0.1% Triton X-100 (Wallac)
- Shaking platform (e.g. Wallac)
- Terbium enhancement solution: 100 mM 4-(2,4,6-trimethoxyphenyl)-pyridine-2,6-dicar-boxylic acid and 1% cetyltrimethylammonium bromide in 1.1 M NaHCO$_3$ (not commercially available; see Section 4.2)

Method

1. Resuspend the washed paramagnetic particles in the wells of the microtitre plate, or immerse the prongs of the manifold support, in 180 μl of europium fluorescence enhancement solution.

---

2. Incubate 10 min on a shaking platform.

3. Record the europium signals in a Delfia Plate Reader Research Fluoro-meter.

4. Next, add 20 μl of a terbium enhancement solution.

5. Shake 10 min.

6. Record the terbium signals.

4. Comments on the OLA protocol

4.1 Design of ligation probes

Oligonucleotides used in this ligation-based assay are complementary to the target sequence, and designed so that the mutation to be analysed lies at the border between two juxtaposed oligonucleotides. One of the oligonucleotides is complementary to a sequence in common for both the normal and mutated sequence. This oligonucleotide is modified at the 5′ end with a phosphate group and at the 3′ end with a biotin residue. The other two oligonucleotides, hybridizing immediately upstream of the first one, are identical in sequence except for their 3′ ends where one is complementary to a normal and the other to a mutant sequence variant. These two allele-specific oligonucleotides are 5′ labelled with detectable groups; that specific for a normal allele carries europium chelates, and the one specific for the mutant, terbium chelates.

4.2 Detection by time-resolved fluorometry

The two lanthanide labels that we use in this assay, chelates of europium and terbium ions, permit sensitive detection of as little as 0.1 μl of amplification reactions and the two colours are well resolved using a commercially available microplate reader. The detectable groups are characterized by an unusual type of fluorescence with a very wide distinction between excitation and emission wavelengths of over 200 nm, and extremely long duration of fluorescence after excitation, permitting time-resolved measurement to discriminate against background. The synthesis of the labelled probes is described by Kwiatkowski *et al.* (16). Probes can also be modified with chelates by reacting oligonucleotides, having reactive amines, with a reagent commercially available from Wallac.

The key component of the fluorescence enhancement solution used for detection of terbium ions is not commercially available. The synthesis is described by Hemmilä *et al.* (17). Other, commercially available substances that can be employed are described by Hemmilä (18). As a further alternative, samarium chelates can be used in place of terbium chelates. The fluorescence of samarium ions can be recorded in the same enhancement solution used for europium, commercially available from Wallac, although samarium ions are detected with approximately fivefold lower sensitivity than europium ions. Other dual-colour labels have been developed for use in the oligonucleotide ligation assay, that can be detected in a regular spectrophotometer (14).

4.3 Properties of the oligonucleotide ligation assay

The method to screen large sets of patient samples for known mutations described here presents several advantages. The ligation reaction accurately distinguishes between sequence variants. If a manifold supports is used to process reactions this further simplifies the procedure and reduces the risk for mistakes. Finally, the dual-label design permits internally controlled analyses of individual samples, enhancing the precision of the analysis. The method is also useful for scoring biallelic genetic variants in forensics or for genetic linkage analysis. Moreover, by selecting an appropriate control sequence, added to the nucleic acid samples at a known concentration, the presence of specific gene sequences can be quantitated using the same ligase-mediated, dual-colour assay format. Recently, we have demonstrated that all components of the ligation reaction may be present in a dried state on the manifold support, reducing the manipulations after PCR to a simple incubation with the manifold support, followed by washes and detection (Samiotaki *et al.* unpublished).

References

1. Besmer, P., Miller, R. C., Caruthers, M. H., Kumar, A., Minamoto, K., van de Sande, J. H., *et al.* (1972). *J. Mol. Biol.*, **72**, 503.
2. Landegren, U., Kaiser, R., Sanders, J., and Hood, L. (1988). *Science*, **241**, 1077.
3. Alves, A. M. and Carr, F. J. (1988). *Nucleic Acids Res.*, **16**, 8723.
4. Wu, D. Y. and Wallace, R. B. (1989). *Gene*, **76**, 245.
5. Barany, F. (1991). *Proc. Natl. Acad. Sci. USA*, **88**, 189.
6. Wu, D. Y. and Wallace, R. B. (1989). *Genomics*, **4**, 560.
7. Barany, F. (1991). *PCR Methods Appl.*, **1**, 5.
8. Landegren, U. (1993). *BioEssays*, **15**, 761.
9. Landegren, U., Samiotaki, M., Kwiatkowski, M. and Nilsson, M. (1996). In *Encyclopedia of molecular biology and molecular medicine* (ed. R. A. Meyer). Vol. 3, p. 391, VCH, Weinheim.
10. Whiteley, N. M., Hunkapiller, M. W., and Glazer, E. N. (1989). US Patent No. 4883750.
11. Luo, J., Bergstrom, D. E., and Barany, F. (1996). *Nucleic Acids Res.*, **24**, 3071.
12. Nickerson, D. H., Kaiser, R., Lappin, S., Stewart, J., Hood, L., and Landegren, U. (1990). *Proc. Natl. Acad. Sci. USA*, **87**, 8923.
13. Samiotaki, M., Kwiatkowski, M., Parik, J., and Landegren, U. (1994). *Genomics*, **20**, 238.
14. Tobe, V. O., Taylor, S. L., and Nickerson, D. A. (1996). *Nucleic Acids Res.*, **24**, 3728.
15. Parik, J., Kwiatkowski, M., Lagerkvist, A., Lagerström Fermér, M., Samiotaki, M., *et al.* (1993). *Anal. Biochem.*, **211**, 144.

16. Kwiatkowski, M., Samiotaki, M., Hurskainen, U., and Landegren, U. (1994). *Nucleic Acids Res.*, **22**, 2604.
17. Hemmilä, I., Mukkala, V-M., Latva, M., and Kiilholma, P. (1993). *J. Biochem. Biophys. Methods*, **26**, 283.
18. Hemmilä, I. (1985). *Anal. Chem.*, **57**, 1676.

9

Detection of sequence variation using primer extension

A.-C. SYVÄNEN

1. Introduction

Most of the currently used methods for detecting single nucleotide polymor-phisms or point mutations are based on amplification of the target DNA by the PCR technique or another amplification method, which allows sensitive and specific analysis of the target DNA sequence by a variety of methods. Pre-viously known sequence variants are distinguished in the amplified fragments by hybridization with sequence-specific oligonucleotide (SSO) probes (see chapter 7) or with the aid of nucleic acid modifying enzymes, such as restric-tion enzymes, DNA ligases, or DNA polymerases (see chapters 8 and 9).

The nucleotide incorporation reaction catalysed by the DNA polymerases is highly sequence-specific. This reaction has been utilized to devise single nucleotide primer extension assays for distinguishing between sequence vari-ants (1–3). The assays are based on detection of the variable nucleotide(s) by specific extension of a primer that anneals immediately adjacent to the variable site with a single labelled nucleotide that is complementary to the nucleotide to be detected using a DNA polymerase. A major advantage of the single nucleotide primer extension principle over hybridization with SSO probes is that the discrimination between the two sequences is based on the sequence-specificity of the reaction catalysed by the DNA polymerase, instead of on differences in thermal stability of the hybrids formed with the SSO probes. Therefore primer extension assays are robust, and insensitive to variations in the reaction conditions, and the same reaction conditions can be employed for detecting any variable nucleotide irrespective of the nucleotide sequence flanking the variable site.

For large scale use a method for detecting single nucleotide polymorphisms should consist of simple steps, including the sample treatment, assay format, and interpretation of the result to facilitate its automation. Different reaction principles for distinguishing between nucleotide sequence variants can be combined with solid phase formats into assays that are more amenable to automation, and easier to combine with non-radioactive detection than methods that depend on gel electrophoretic separation steps (for a review, see ref. 4).

The present chapter describes the solid phase minisequencing method, which is based on a single nucleotide primer extension reaction on a solid support. The method has been used for detecting numerous mutations causing human genetic disorders (5), and for analysing biallelic variation in gene mapping studies, and for the identification of individuals (6). It has also proven to be a valuable tool for quantitative PCR analysis of mRNA and DNA (7, 8). The minisequencing method has been adapted to colorimetric detection (9) and a system for multiplex, fluorescent typing of HLA has been devised (10).

2. Principle of the solid phase minisequencing method

In the solid phase minisequencing method a DNA fragment spanning the variable or mutant nucleotide position is first amplified by PCR using one biotinylated and one non-biotinylated PCR primer. The amplified fragment carrying a biotin residue in the 5′ end of one of its strands is affinity-captured on an avidin- or streptavidin-coated solid support. The excess of non-biotinylated primer and nucleoside triphosphates (dNTPs) from PCR are removed by washing the solid support, and the captured DNA fragment is rendered single-stranded by alkaline denaturation. The nucleotides at a variable or mutant site are identified in the captured DNA fragment by single nucleotide primer extension reactions. In this 'minisequencing' reaction the 3′ end of a detection step primer designed to anneal immediately adjacent to the variable nucleotide position is extended by a DNA polymerase with one labelled nucleoside triphosphate complementary to the nucleotide at the variable site. The incorporated label is measured, and it serves as a highly specific indicator of the nucleotide present at the variable or mutant site. *Figure 1* illustrates the principle of the solid phase minisequencing method.

3. Design and synthesis of primers

A prerequisite for setting-up the solid phase minisequencing method is that nucleotide sequence information is available for designing primers for PCR and for the minisequencing primer extension reaction. PCR primers and one minisequencing detection primer per variable nucleotide position are required. During the amplification, the PCR primers should produce fragments, preferably between 50 and 500 base pairs in size, containing the variable nucleotide positions. The PCR primers should be 20–23 nucleotides long, have similar melting temperatures, and non-complementary 3′ ends (11). One of the PCR primers is biotinylated in its 5′ end during its synthesis using a biotinyl phosphoramidite reagent (e.g. RPN 2012 from Amersham). The biotinylated primer can be used without purification given that the efficiency

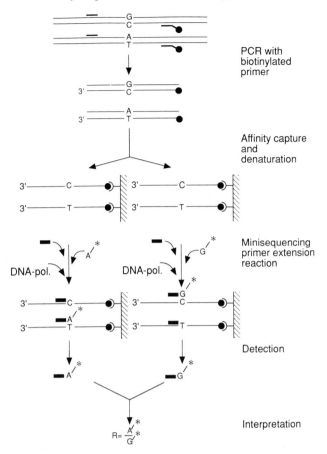

Figure 1. Principle and steps of the solid phase minisequencing method.

of the biotinylation reaction on the oligonucleotide synthesizer has been close to 90%.

The minisequencing detection step primer is designed to be complementary to the biotinylated strand of the PCR product immediately 3′ of the variable nucleotide position (*Figure 2A*) It should preferably be 20 nucleotides long and, to ensure that possible non-specific PCR products remain undetected, it should have at least five nucleotides that do not overlap with the non-biotinylated PCR primer (*Figure 2A*). Deletions (insertions) can be detected analogously to point mutations using a primer that anneals immediately adjacent to the deletion (insertion) breakpoint. In the normal allele the first nucleotide within the deletion, and in the mutant allele the first nucleotide following the deletion, are detected in the minisequencing reactions (*Figure 2B, C*). For small deletions of a few nucleotides the PCR is carried out as for point mutations (*Figure 2B*), but for detecting larger deletions/insertions two biotinylated

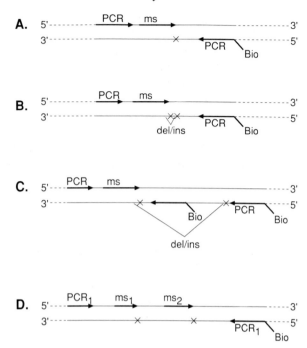

Figure 2. Strategy for designing primers for the solid phase minisequencing method. The amplification primers are denoted 'PCR', the minisequencing primers are denoted 'ms', and 'Bio' is biotin. The nucleotides to be identified in the minisequencing reactions are indicated by 'X'. (A) Primers for detecting a single nucleotide variation. The minisequencing primer is complementary to the biotinylated strand of the amplified fragment. (B) Primers for detecting small deletions or insertions of a few nucleotides. The deletion/insertion is indicated by 'del/ins'. (C) Primers for detecting large deletions or insertions. Two biotinylated PCR primers, one of which is located within the delection/insertion and the other outside the deletion/insertion are used. (D) Multiplex minisequencing exemplified by primers for the identification of polymorphic nucleotides at four sites within two different PCR products (PCR_1 and PCR_2) using minisequencing primers (ms_1–ms_4) of different size.

PCR primers are used to give PCR products of similar size what are amplified with equal efficiency from both alleles (*Figure 2C*).

4. Guidelines for PCR amplification

Various types of DNA samples, treated as is suitable for PCR amplification, can be analysed (12). RNA samples are amplified after synthesis of a first strand cDNA by reverse transcriptase (13). PCR is carried out according to a standard protocol (14) most conveniently in a 50–100 μl reaction volume, except that when streptavidin-coated microtitre plate wells are to be used as solid support for the minisequencing reaction, the total concentration of the biotinylated PCR primer(s) should not be higher than 0.2 μM (10 pmol per 50 μl reaction volume) in order not to exceed the biotin binding capacity of the wells (see *Protocol 1*). Precautions for avoiding DNA contaminations in the PCR should be taken (15). The PCR should be optimized to be efficient and specific. One-tenth of the PCR products should be clearly visible as a single band in an agarose gel after staining with ethidium bromide.

5. The solid phase minisequencing method in practice

In our basic protocol for performing the solid phase minisequencing assay, [³H]dNTPs are used as labels and streptavidin-coated microtitre plate wells serve as the solid support. The protocol given below is applicable without changes for detecting any single nucleotide variation, small deletion, or insertion. A modification of the protocol, in which avidin-coated polystyrene microparticles with higher biotin binding capacity than microtitre plate wells are used as solid support, is also presented.

5.1 Affinity capture

After amplification, the PCR products are captured in streptavidin-coated microtitre plate wells. Immobilization of the PCR products facilitates efficient removal of the excess of nonbiotinylated PCR primer and dNTPs, as well as denaturation of the PCR product. It is important for the specificity of the minisequencing reaction that all dNTPs from the PCR are completely removed by the washing steps because other dNTPs than the single labelled one that are present during the minisequencing reaction will cause non-specific extension of the detection step primer. The use of an automatic micotitre plate washer (*Protocol 1*) saves times and labour, and improves the washing efficiency.

Protocol 1.   Affinity capture and denaturation of the 5′ biotinylated PCR products in streptavidin-coated microtitre plate wells

Equipment and reagents

- Incubator or water-bath with shaker at 37 °C
- Multichannel pipettes and/or microtitre plate washer (optional)
- Streptavidin-coated microtitre plates (Combiplate 8, Labsystems)
- 50 mM NaOH

- PBS–Tween buffer: 20 mM sodium phosphate buffer pH 7.5, 100 mM NaCl, 0.1% (v/v) Tween 20
- TENT buffer: 40 mM Tris–HCl pH 8.8, 1 mM EDTA, 50 mM NaCl, 0.1% Tween 20

Protocol 1. *Continued*

Method

1. Transfer two 10 μl aliquots (or four 10 μl aliquots for duplicate assays) of each PCR product to streptavidin-coated microtitre plate wells. Include two negative controls containing PCR buffer only.

2. Add 40 μl of PBS–Tween buffer to each well and seal the wells with a sticker.

3. Incubate the microtitre plate for 1.5 h at 37 °C with gentle shaking. Discard the contents of the wells.

4. Wash the wells manually three times at room temperature (about 20 °C) by adding 200 μl of TENT buffer. Empty the wells thoroughly between the washes by tapping the plates upside down against a tissue paper. Alternatively, wash the wells five times with 300 μl of TENT buffer in a microtitre plate washer.

5. Add 100 μl of 50 mM NaOH to each well.

6. Incubate at room temperature for 2–5 min. Discard the NaOH, and wash as above in step 4.

7. Continue with the minisequencing reactions as described in *Protocol 3*.

The biotin binding capacity of the microtitre plate wells sets an upper limit to the amount of biotinylated PCR product (and excess of biotinylated primer) that can be present during the capturing reaction. The biotin binding capacity of the wells given in *Protocol 1* is 2–5 pmol of biotinylated oligonucleotide. Therefore we use the biotinylated primer at 0.2 μM (0.2 pmol/μl) concentration and analyse 10 μl of the PCR product per well. If the biotin binding capacity of the wells has been exceeded, low signals or lack of signals will result.

Alternatively, if higher biotin binding capacity is required, e.g. for multiplex PCR applications, another affinity matrix with high biotin binding capacity such as avidin-coated polystyrene microparticles (Fluoricon assay particles, 0.7–0.9 μM, IDEXX Corp.) with extremely high biotin binding capacity (>2 nmol/mg of particles) (see *Protocol 2*), or streptavidin-coated magnetic polystyrene beads (Dynabeads M-280, Dynal, binding capacity 150–300 pmol of biotinylated primer/mg of beads) can be used (16).

5.2 The minisequencing primer extension reaction

The variable nucleotide(s) are detected in the immobilized single-stranded DNA fragment by extension of the detection primer with a single labelled nucleoside triphosphate. In *Protocols 3* and *4*, a thermostable DNA polymerase is used to incorporate a ^{3}H-labelled dNTP in the minisequencing reaction.

Protocol 2. Affinity capture and denaturation of the 5′ biotinylated PCR products on avidin-coated polystyrene microparticles

Equipment and reagents

- Microcentrifuge
- Avidin-coated polystyrene particles (Fluoricon assay particles, 0.7–0.9 μM, 5% (w/v) suspension; IDEXX Corp.)
- TENT buffer (see *Protocol 1*)
- 50 mM NaOH containing 0.1% Tween 20

Method

1. To wash the polystyrene microparticles before use, transfer 5 μl of the 5% suspension of particles per sample to be analysed to a microcentrifuge tube. Add 1 ml of TENT buffer and suspend the particles by vortexing. Pellet the particles by centrifugation for 2 min at 13 000 g. Resuspend the particles to a 5% suspension in TENT buffer.

2. Transfer 5 μl aliquots of the suspension of washed particles to microcentrifuge tubes. Add two 10–40 μl aliquots of each biotinylated PCR product (or four aliquots for duplicate assays) and TENT buffer to a final volume of 100 μl to the microcentrifuge tubes. Include two negative controls without PCR product.

3. Incubate the samples at room temperature for 15 min. Pellet the particles by centrifuging for 2 min at 13 000 g and discard the supernatant.

4. Wash the particles three times by adding 1 ml of TENT buffer, followed by vigorous vortexing for 10 sec. Pellet the particles by centrifugation for 2 min at 13 000 g and discard the supernatant.

5. Add 100 μl of 50 mM NaOH, 0.1% Tween 20 to the tubes and suspend the particles thoroughly.

6. Incubate the samples at room temperature for 2–5 min. Pellet the particles as in step 3, discard the NaOH solution, and wash as in step 4.

7. Continue with the minisequencing reactions as described in *Protocol 4*.

The reaction conditions for annealing the primer to the immobilized DNA strand are non-stringent. Therefore the same reaction conditions can be used for analysis of any DNA fragment, irrespective of the nucleotide sequence of the detection step primer. An advantage of using a thermostable polymerase in the minisequencing primer extension reaction is that annealing of the primer to the template and extension of the primer with the labelled dNTP can be carried out simultaneously at a fairly high temperature (50 °C) that is favourable for both reactions.

Protocol 3. The minisequencing reactions in microtitre plate wells using ^{3}H-labelled dNTPs

Equipment and reagents

- Incubator or water-bath at 50 °C
- Multichannel pipettes and/or microtitre plate washer (optional)
- Liquid scintillation counter
- Minisequencing detection primer (5 μM)
- 10 × concentrated DNA polymerase buffer: 500 mM Tris–HCl pH 8.8, 150 mM $(NH_4)_2SO_4$, 15 mM $MgCl_2$, 1% (v/v) Tween 20, 0.1% (w/v) gelatin (the buffer supplied with the DNA polymerase enzyme can also be used)

- Thermostable DNA polymerase: *Taq* DNA polymerase (Promega Biotech) or Dyna-zyme™ DNA polymerase (Finnzymes)
- ^{3}H-labelled deoxynucleoside triphosphates: [^{3}H]dATP, TRK 633; [^{3}H]dCTP, TRK 625; [^{3}H]dGTP, TRK 627; [^{3}H]dTTP, TRK 576 (Amersham)
- TENT buffer (see *Protocol 1*)
- 50 mM NaOH
- Scintillation fluid (e.g. HI-Safe II, Wallac)

Method

1. Prepare separate master mixtures for each nucleotide to be detected by combining 5 μl of 10 × DNA polymerase buffer, 2 μl of 5 μM detection step primer (10 pmol), 0.1 μCi (usually 0.1 μl) of the [^{3}H]dNTP complementary to the nucleotide to be detected, 0.1 U of DNA polymerase, and distilled water to 50 μl per reaction. The reaction mixtures can be prepared during the capturing reaction and they can be stored at room temperature until use.

2. Add 50 μl of reaction mixture to the microtitre plate wells carrying the amplified, denatured samples (*Protocol 1*). Seal the wells with a sticker.

3. Incubate the plate for 10 min at 50 °C. Discard the contents of the wells and wash the wells as described in *Protocol 1*, step 4.

4. Add 60 μl of 50 mM NaOH to each well and incubate at room temperature for 2–5 min.

5. Transfer the 50 mM NaOH solution containing the eluted primer to scintillation vials, add scintillation fluid, and measure the eluted ^{3}H in a liquid scintillation counter.

If a scintillation counter for microtitre plates is available, streptavidin-coated microtitre plates manufactured from scintillating plastic (ScintiStrips™, Wallac) can be used as the solid phase (17). This will simplify the procedure in that the final washing and denaturing steps and the transfer of the eluted primer to scintillation vials can be omitted.

The use of avidin-coated microparticles as solid support gives the advantage of an almost unlimited biotin binding capacity, but introduces the drawback of requiring centrifugation during the washing steps. Except for the handling of the particles, the minisequencing reactions conditions given in

Protocol 4 are identical to those in Protocol 3 for the microtitre plate format of the assay.

Protocol 4. The minisequencing reactions on avidin-coated polystyrene microparticles using ^{3}H-labelled dNTPs

Equipment and reagents
- Microcentrifuge
- See *Protocol 3*
- 50 mM NaOH, 0.1% Tween 20

Method

1. Prepare separate master mixtures for each nucleotide to be detected by combining 5 μl of 10 × DNA polymerase buffer, 2 μl of 5 μM detection step primer (10 pmol), 0.1 μCi (usually 0.1 μl) of the [^{3}H]dNTP complementary to the nucleotide to be detected, 0.1 U of DNA polymerase, and distilled water to 50 μl per reaction.

2. Add 50μl of reaction mixture to the pelleted microparticles carrying the amplified, denatured samples (*Protocol 2*), and resuspend the pellet by vortexing briefly.

3. Incubate the samples for 10 min at 50°C. Pellet the particles by centrifugation for 2 min at 13000 g and discard the supernatant. Wash the particles three times as described in *Protocol 2*, step 4.

4. Add 100 μl of 50 mM NaOH, 0.1% Tween 20 to each well and incubate at room temperature for 2–5 min.

5. Transfer 90 μl of the 50 mM NaOH, 0.1% Tween 20 solution containing the eluted primer to scintillation vials, add scintillation fluid, and measure the eluted ^{3}H in a liquid scintillation counter.

5.3 Interpretation of the result

In samples from individuals homozygous for the analysed nucleotide a signal will be generated in one of the minisequencing reactions, and in samples from heterozygous individuals signals will be generated in both reactions. The result of the assay is obtained as a numeric c.p.m. (counts per minute) value corresponding to the amount of [^{3}H]dNTP incorporated. When the assay has been successful, positive signals will be at least 1000 c.p.m., and the background should be below 100 c.p.m. The ratio (R-value) between the c.p.m. values obtained in the two reactions defines the genotype of the analysed samples. *Table 1* shows as an example the results from analysing a polymorphic nucleotide in the cathecol-*O*-methyltransferase gene (18). Calculation of the R-value eliminates variations in the amount of incorporated [^{3}H]dNTPs due

Table 1. Result from detection of the polymorphic nucleotide (A or G) in codon 158 of the catechol-*O*-methyltransferase gene by solid phase minisequencing[a]

Genotype of sample	Incorporated radioactivity (c.p.m.)[b]		Ratio [³H]dATP/[³H]dGTP
	[³H]dATP	[³H]dGTP	
Homozygous (AA)	3200	53	60
Homozygous (AA)	52	1580	0.033
Heterozygous (AG)	2819	1390	2.0
No DNA	76	50	–

[a] Primer sequences for the assay are given in ref. 18.
[b] The specific activity of [³H]dATP was 73 Ci/mmol and that of [³H]dGTP was 36 Ci/mmol.

to variation between samples in the efficiency of PCR. The R-value will usually be > 10 or < 0.1 in samples from subjects homozygous for a variable nucleotide and between 0.5 and 2 in samples from heterozygous subjects, depending on the specific activities of the [³H]dNTPs used. If the sequence contains one (or more) identical nucleotides immediately next to the nucleotide at the variable site, one (or more) additional [³H]dNTP will be incorporated in the minisequencing reaction, which obviously will affect the R-value.

Despite the fact that ³H is a radionucleotide, [³H]dNTPs are convenient as labels because ³H has a long half-life (13 years) and is a weak beta-emitter. Because the result from the beta counter is obtained as a numeric value directly corresponding to the amount of [³H]dNTPs incorporated in the minisequencing reaction, they are a most advantageous alternative as labels for quantitative applications.

6. Quantitative PCR analysis by solid phase minisequencing

The R-value obtained in the solid phase minisequencing method reflects the ratio between two sequence variants when they are present in a sample as a mixture in any other ratio than that in samples from homozygous subjects (allele ratio 2:0) or heterozygous subjects (allele ratio 1:1). Since the two sequences are identical, with the exception of a single nucleotide, they can be assumed to be amplified with equal efficiency during PCR, irrespective of whether the amplification is in its exponential phase. The initial ratio between the two sequences in a sample can be calculated from the R-value by taking into account the specific activities and the number of [³H]dNTPs incorporated in the minisequencing reactions. Both these factors affecting the R-value are known in advance and can easily be accounted for.

Instead of calculating the initial ratio from the known specific activities of the [³H]dNTPs used in the minisequencing reaction, the ratio between two sequences can be determined by comparing the obtained R-value with a standard curve prepared by analysing mixtures of known amounts of the corresponding two sequences (8). The use of a standard curve corrects for differences in specific activity and the number of [³H]dNTPs incorporated, but also for a possible small misincorporation of [³H]dNTP by the DNA polymerase, which may affect the result, particularly when a sequence present as a small minority of a sample is to be quantified. The high specificity of the single nucleotide incorporation catalysed by the DNA polymerase allows detection of one sequence present as a small minority ($< 1\%$) in a sample (5, 19). To quantify a sequence present as a small minority of the sample, it is advantageous to use a solid support with high biotin binding capacity, such as avidin-coated polystyrene particles (see *Protocol 2*) in the capturing reaction.

Quantitative PCR analysis of mixed samples by the solid phase mini-sequencing method has been used for determination of the population frequencies of disease-causing mutant alleles and polymorphisms from large pooled DNA samples (5, 6), for determination of the proportion of hetero-plasmic mutations of the mitochondrial DNA (8) or mutant blast cells in bone marrow samples (19), and for comparing transcript levels of highly homologous genes (20). *Table 2* shows an example of the results from comparing the levels of expression of the highly homologous fibrillin 1 (*FBN1*) and fibrillin 2 (*FBN2*) genes in cultured fibroblasts derived from patients with the Marfan syndrome (MFS) and from control individuals. As can be seen, the relative level of expression of the *FBN1* as compared to that of the *FBN2* was significantly elevated in the fibroblasts from two of the patients. *Table 2* also

Table 2. Result from determination of the relative amounts of fibrillin 1 and fibrillin 2 mRNA by solid phase minisequencing[a]

Sample	Incorporated radioactivity (c.p.m.)[b]		Ratio	
	[³H]dCTP	[³H]dTTP	[³H]dCTP/[³H]dTTP	FBN1/FBN2 mRNA[c]
MFS[d] patient 1	2750 ± 244	92 ± 20	27.9	52
MFS patient 2	3222 ± 233	4569 ± 545	0.70	1.3
MFS patient 3	4945 ± 359	764 ± 97	6.47	12
Control 1	2024 ± 157	2527 ± 118	0.80	1.5
Control 2	4047 ± 217	3672 ± 363	1.10	2.1
Control 3	1941 ± 33	1081 ± 15	1.80	3.4

[a] The data are from ref. 20.
[b] Mean values of four parallel assays from the same PCR reaction ± standard deviations.
[c] Ratio obtained by correcting for the specific activities of the [³H]dNTPs; [³H]dCTP = 40 Ci/mmol, [³H]dTTP = 121 Ci/mmol.
[d] MFS = Marfan syndrome.

illustrates the good reproducibility of the minisequencing method. In most of the samples the coefficient of variation (CV) between quadruplicate samples is below 10%.

In many applications, such as those mentioned above, it is sufficient to determine the relative amount of two sequences that are present as a mixture in a sample, and thus serve as internal PCR standards for each other. To determine the absolute amount of a sequence present in a sample, a known amount of a standard sequence differing from the sequence to be quantified by a single nucleotide is added to the sample before the PCR amplification (7, 21) to serve as an internal standard. Depending on the application, either synthetic oligonucleotides (8), genomic DNA of known genotype (5), or RNA prepared by *in vitro* transcription (7) can be used as standard sequences. The R-value determined in the minisequencing assay then allows calculation of the absolute amount of the sequence present in the analysed sample. When the absolute amount of a sequence is determined, the size of the analysed sample must be defined, e.g. in terms of number of cells, total amount of protein, or RNA.

7. Minisequencing with colorimetric detection

Nucleoside triphosphates modified with haptens can also be used as detectable groups in the minisequencing reaction, provided that they are incorporated sequence-specifically by a DNA polymerase. The incorporated haptens can then be detected indirectly with the aid of antibodies conjugated to alkaline phosphatase or peroxidase using colorimetric or chemiluminescent substrates. The concept of using hapten labelled nucleotides in the solid phase minisequencing assay has been presented in two early studies (1, 9), but colorimetric detection has not become practically feasible until all four nucleotides have become available as hapten-modified analogues detectable with antibody–enzyme conjugates (22). When streptavidin-coated microtitre plate wells are used as the solid support, the format of the minisequencing assay with haptens as detectable groups is identical to that of enzyme immunoassays. The affinity capture step prior to the colorimetric detection procedure is carried out according to *Protocol 1*. *Protocol 5* describes a solid phase minisequencing assay adapted for use with fluorescein (FITC) labelled ddNTPs in conjunction with ThermoSequenase™ DNA polymerase. Anti-FITC antibodies conjugated to alkaline phosphatase and a chromogenic substrate are used as the final detection method.

Table 3 illustrates the results obtained when three samples of different genotype were analysed for the G to A transition in codon 506 of the coagulation factor V gene (23, 24) by the colorimetric minisequencing method. As with [^{3}H]dNTPs as labels, the R-values ($G_{405\ nm}/A_{405\ nm}$) clearly define the genotype of the samples.

Protocol 5. The minisequencing reactions with colorimetric detection

Equipment and reagents

- Incubator or water-bath at 37 °C and 50 °C
- Multichannel pipettes and/or microtitre plate washer (optional)
- Spectrophotometric reader for microtitre plate
- ThermoSequenase™ DNA polymerase (Amersham, E79000Y)
- Minisequencing detection primer (5 μM)
- 10 × ThermoSequenase buffer: 260 mM Tris–HCl pH 9.5, 65 mM MgCl₂
- FITC labelled dideoxynucleoside triphosphates (Renaissance Nucleotide Analogues, NEN/DuPont); F-ddATP, NEL-402; F-ddCTP, NEL-400; F-ddGTP, NEL-403; F-ddUTP, NEL-401

- Dideoxynucleoside triphosphates: 5 mM solutions (Pharmacia Biotechnology)
- TENT buffer (see *Protocol 1*)
- 50 mM NaOH
- Anti-FITC–alkaline phosphatase (AP) conjugate (Boehringer Mannheim, 1426338)
- TBS: 25 mM Tris–HCl pH 7.5, 0.15 M NaCl
- Conjugate buffer: 0.1% Tween 20, 1% bovine serum albumin in TBS
- TBS–Tween buffer; 0.1% Tween 20 in TBS
- Alkaline buffer: 1 mM diethanolamine, 0.5 mM MgCl₂ pH 10
- 2 mM 4-nitrophenyl phosphate in the alkaline buffer

Method

1. Prepare separate master mixtures for each nucleotide to be detected by combining 5 μl of 10 × ThermoSequenase buffer, 2 μl of 5 μM detection step primer (10 pmol), 5 μl of a 3.5 μM solution (17.5 pmol) of the F-ddNTP complementary to the nucleotide to be detected, 5 μl of a 5 μM solution of the three other unlabelled ddNTPs, 0.12 U of ThermoSequenase™ DNA polymerase, and distilled water to 50 μl per reaction.

2. Add 50 μl of reaction mixture to the microtitre plate wells carrying the amplified, denatured samples. Seal the wells with a sticker.

3. Incubate the plate for 15 min at 50 °C. Discard the contents of the wells and wash the wells with TENT as described in *Protocol 1*, step 4.

4. Add 60 μl of anti-FITC–AP conjugate diluted 1:1000[a] in conjugate buffer to the well.

5. Incubate the plate for 2 h at 37 °C. Discard the contents of the wells and wash the wells three times with TBS–Tween buffer as described in *Protocol 1*, step 4, and once with the alkaline buffer.

6. Add 160 μl of 2 mM 4-nitrophenyl phosphate in the alkaline buffer to the wells.

7. Incubate the plate for 20 min at room temperature.

8. Measure the absorbance of the formed coloured end-product at 405 nm.

---

[a]The dilution of each different batch of anti-FITC–AP conjugate may have to be titrated to obtain optimal signal-to-noise ratios.

Table 3. Result from detection of the G to A transition in codon 506 of the blood coagulation factor V gene by colorimetric minisequencing[a]

Genotype of sample	Absorbance of 405 nm		Ratio[b] $G_{405 nm}/A_{405 nm}$
	G-reaction	**A-reaction**	
Normal (GG)	0.904	0.075	11 ± 1.2
	0.915	0.091	
Heterozygous (AG)	0.942	1.019	0.94 ± 0.017
	0.969	1.006	
Homozygous (AA)	0.288	1.475	0.18 ± 0.028
	0.223	1.376	

[a] Data provided by Progene Lab AB, Uppsala, Sweden (ref. 24).
[b] Mean values of the four G/A ratios $\pm$ standard deviation.

It would be more convenient if the indirect immunochemical detection procedure used to detect the incorporated FITC labelled ddNTPs could be substituted with direct fluorescence detection. Unfortunately, the detection sensitivity of presently available standard fluorescence readers for microtitre plates is not sufficient for reliable direct detection of the incorporated F-ddNTPs.

8. Multiplex, fluorescent minisequencing

DNA sequencing instruments offer the possibility for sensitive detection of minisequencing primers extended with fluorescent ddNTPs. Since the primers are separated by size during the electrophoresis on the DNA sequencer, multiple polymorphic sites can be analysed simultaneously in an undivided sample using primers varying in size. The size of the extended primer defines the position of the polymorphism, and the incorporated F-ddNTP gives the identity of the nucleotide at each site (*Figure 2D*).

In the multiplex, fluorescent minisequencing method described in *Protocols 6–8*, multiple primers differing in size by three nucleotides are extended by FITC labelled ddNTPs and detected using the ALF DNA sequencer. Because the ALF instrument is based on the use of a single label, separate reactions must be performed for each of the four nucleotides. The use of avidin-coated comb-shaped manifold supports (25) designed to fit the slots of the sequencing gel of the ALF instrument allows practical handling of large numbers of samples. We have set-up the method for multiplex genotyping of the HLA-DQA1 and DRB1 genes at nine polymorphic sites (10), but the method is generally applicable for detecting any single nucleotide polymorphisms or point mutations.

Protocol 6. Affinity capture and denaturation of the 5′ biotinylated PCR products on avidin-coated manifold supports

Equipment and reagents

- Avidin-coated comb-shaped manifold supports (Autoload, Pharmacia Biotechnology)
- 10-well reaction plates accommodating four teeth of the manifold support per well (Autoload, Pharmacia Biotechnology)
- Incubator at 37 °C
- 100 mM NaOH

- 4 × capturing buffer: 4 M NaCl, 20 mM Tris–HCl pH 7.5, 2 mM EDTA
- Rinsing buffer: 2 M NaCl, 10 mM Tris–HCl pH 7.5, 1 mM EDTA
- TE buffer: 10 mM Tris–HCl pH 8.0, 1 mM EDTA

Method

1. Transfer 15 µl aliquots of each of the 5′ biotinylated PCR products to wells of the reaction plates. Add 30 µl of 4 × capturing buffer and distilled water to a final volume of 120 µl per well.

2. Insert four teeth of the manifold support per reaction plate well. Place the plate on a wet tissue paper and cover the plate and manifolds with plastic wrap, or use as an alternative a humid chamber.

3. Incubate the plates for 2–24 h at 37 °C. It is important that the wells do not dry during the incubation period.

4. Rinse the manifold support in a vial containing rinsing buffer.

5. Transfer the manifold support to another reaction plate containing 150 µl of 100 mM NaOH per well.

6. Incubate the plates for 5 min at room temperature (about 20 °C).

7. Rinse the manifold support vigorously once in a vial containing TE and once in a vial containing distilled water.

8. Continue with the minisequencing reactions as described in *Protocol 7*.

We have estimated that the biotin-binding capacity of the manifold support in combination with the high detection sensitivity of the sequencing instrument permit capture and analysis of 20–30 5′ biotinylated PCR products. If more than six different PCR products are to be analysed, the volume of PCR product added to the capturing reaction recommended in *Protocol 6*, is reduced, and obviously, a more concentrated capturing buffer can be used to make space for more PCR products in the wells of the reaction plates. Other affinity matrices, such as the avidin-coated microparticles employed in *Protocol 2*, can be used instead of the manifolds as supports for the multiplex, fluorescent minisequencing reactions.

Protocol 7. The multiplex, fluorescent minisequencing reactions on avidin-coated manifold supports

Equipment and reagents

- 40-well reaction plates accommodating four teeth of the manifold support per well (Autoload, Pharmacia Biotechnology)
- Incubator or water-bath at 50 °C (heat blocks shaped for the manifold reaction plates are available from Pharmacia Biotechnology)
- Minisequencing primers of different size (15 μM solutions)

- ThermoSequenase™ DNA polymerase (see *Protocol 5*)
- 10 × ThermoSequenase buffer (see *Protocol 5*)
- FITC labelled dideoxynucleoside triphosphates (see *Protocol 5*)
- Dideoxynucleoside triphosphates (see *Protocol 5*)

Method

1. Prepare four master mixtures for the multiplex minisequencing reactions by combining 2 μl of 10 × ThermoSequenase buffer, 2 μl of a 15 μM solution of each detection step primer (30 pmol), 5 μl of a 1.5 μM solution (7.5 pmol) of one F-ddNTP, 5 μl of a 5 μM solution of the three other unlabelled ddNTPs, 0.12 U of ThermoSequenase™ DNA polymerase, and distilled water to a final volume of 20 μl per reaction.

2. Transfer 20 μl aliquots of the four reaction mixtures to four adjacent reaction plate wells.

3. Insert the manifold support carrying the denatured PCR products (*Protocol 6*) into the wells of the reaction plates. Incubate the plates for 15 min at 50 °C.

4. Place the plates on ice if the electrophoresis on the DNA sequencer will be performed during the same day (*Protocol 8*). Store the plates at −20 °C for longer periods.

Fluorescent multiplex minisequencing methods, in which the primers are extended using ddNTPs labelled with four different fluorophores followed by multicolour detection in one lane using an Applied Biosystems DNA sequencer, have been set-up for detection of mutations in the *HPRT* gene (26) and for genotyping of mitochondrial DNA (27). In principle this approach reduces the number of multiplex minisequencing reactions per sample from four to one. Because separation of primers extended with four fluorophores with different electrophoretic mobility in one lane requires longer running times, the theoretical fourfold increase in capacity compared to our ALF-based system, in which the gels can be rerun up to six times, is not fully achieved.

Protocol 8. Fluorescence analysis of the extended minisequencing primers

Equipment and reagents

- Fluorescent DNA sequencing instrument (ALF, Pharmacia Biotechnology)
- Glass plates, spacers, and combs for short (18 × 27 cm) gels for the ALF sequencer
- ALF Fragment Manager software (Pharmacia Biotechnology)
- Urea

- 50% Hydrolink polyacrylamide (Long Ranger, AT Biochem)
- 5 × TBE: 445 mM Tris base, 445 mM boric acid, 10 mM EDTA pH 8.0
- 100% deionized formamide containing 5% mg/ml of Dextran Blue 2000

Method

1. Prepare a 18 × 27 cm polyacrylamide gel containing 10% Hydrolink polyacrylamide, 8 M urea in 1.2 × TBE according to the instructions of the ALF DNA sequencer.

2. Use 1.2 × TBE as running buffer. Pre-heat the gel to 44 °C.

3. Rinse the manifold supports carrying the products of the multiplex fluorescent minisequencing reactions (*Protocol 7*) in 1.2 × TBE, and insert the manifold support into the slots of the gel.

4. Incubate for 5 min at 44 °C to release the primers from the immobilized template. Remove the manifold supports.

5. Carry out the electrophoresis for 50–60 min at 1200 V, 58 mA, and 35 W. The same gel can be reloaded up to six times.

6. Interpret the results of the electrophoretic run saved in the computer of the sequencing instrument with the aid of the ALF Fragment Manager software.

In samples from individuals homozygous for the analysed polymorphism only one fluorescence peak is detected at the time point corresponding to the size of the primer, whereas in heterozygous samples there are two peaks at the corresponding time point. *Figure 3* shows an example of the result from typing a sample at six polymorphic sites in exon 2 of the HLA-DQA1 gene by the multiplex, fluorescent minisequencing method (10).

9. Conclusion

Any point mutation or polymorphic nucleotide can be detected by the solid phase minisequencing method employing the same reaction conditions. All reagents and equipment required for the variants of the solid phase minisequencing method described above are generally available from common suppliers of molecular biological reagents. For these reasons the method is

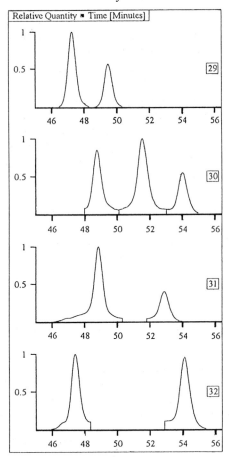

Figure 3. Example of the result from typing the HLA-DQA1 gene by multiplex minise-quencing ine one sample using fluorescence detection on a DNA sequencer. The lanes corresponding to the primers extended with fluorescent ddATP, ddCTP, ddGTP, and ddTTP, respectively, are shown from top to bottom with the relative fluorescence inten-sity on the *y* axis and the separation time in minutes on the *x* axis. The peaks correspond-ing to the extended primers are detected according to their size ranging from 18 nucleotides to 33 nucleotides. The sequence pattern (A/T) (C/G) A C G (C/T) of the sample corresponds to the HLA-DQA1 genotype 0101/0103. The data are from ref. 10.

easy to set-up both in the research laboratory and for routine diagnostics for detection of previously known point mutations and single nucleotide poly-morphisms of interest. The microtitre plate format of the assay can be auto-mated by pipetting robots, and the manifold format of the assay involves extremely simple handling procedures and automatic interpretation of the result. The method is, however, fairly expensive to use when very large numbers of samples are to be analysed simultaneously for multiple mutations or polymorphisms. At present an intense development of methods for detecting point mutations is going on. It can be foreseen that in the near future methods for detecting mutations and sequence variants that involve more advanced production techniques than the current methods, but will be less labour-intensive and more cost-effective in large scale analyses will become available.

A promising approach towards this goal is the development of methods based on arrays of oligonucleotides on miniaturized solid supports. These so-called DNA chips carrying large numbers of SSO probes, originally intended for DNA sequencing by hybridization (28, 29), will be powerful tools for the simultaneous detection of large numbers of previously known mutations and polymorphisms. Analogously, oligonucleotide arrays carrying large numbers of immobilized minisequencing primers can serve as solid supports in assays based on the single nucleotide primer extension reaction (26, 30). In this format the advantage of the robust minisequencing reaction is combined with the high capacity of oligonucleotide arrays as solid supports into highly specific and efficient tests for single nucleotide variations.

References

1. Syvänen, A-C., Aalto-Setälä, K., Harju, L., Kontula, K., and Söderlund, H. (1990). *Genomics*, **8**, 684.
2. Sokolov, B. P. (1990). *Nucleic Acids Res.*, **18**, 3671.
3. Kuppuswami, M. N., Hoffmann, J. W., Kasper, C. K., Spitzer, S. G., Groce, S. L., and Bajaj, S. P. (1991). *Proc. Natl. Acad. Sci. USA*, **88**, 1143.
4. Syvänen, A-C. and Landegren, U. (1994). *Hum. Mutat.* **3**, 172.
5. Syvänen, A-C., Ikonen, E., Manninen, T., Bengström, M., Söderlund, H., Aula, P., et al. (1992). *Genomics*, **12**, 590.
6. Syvänen, A-C., Sajantila, A., and Lukka, M. (1993). *Am. J. Hum. Genet.*, **52**, 46.
7. Ikonen, E., Manninen, T., Peltonen, L., and Syvänen, A-C. (1992). *PCR Methods Appl.*, **1**, 234.
8. Suomalainen, A., Kollmann, P., Octave, J-N., Söderlund, H., and Syvänen, A-C. (1993). *Eur. J. Hum. Genet.*, **1**, 88.
9. Harju, L., Weber, T., Alexandrova, L., Lukin, M., Ranki, M., and Jalanko, A. (1993). *Clin. Chem.*, **39**, 2282.
10. Pastinen, T., Partanen, J., and Syvänen, A-C. (1996). *Clin. Chem.*, **42**, 1391
11. Dieffenbach, C. W., Lowe, T. M. J., and Dveksler, G. S. (1993). *PCR Methods Appl.*, **3**, S30.
12. Higuchi, R. (1989). In *PCR technology. Principles and applications* (ed. H. A. Erlich), p. 31. Stockton, New York.
13. Kawasaki, E. S. (1990). In *PCR protocols: a guide to methods and applications* (ed. M. A. Innis, D. H. Gelfand, J. J. Sninsky, and T. J. White), p. 21. Academic Press, San Diego.
14. Innis, M. A. and Gelfand, D. H. (1990). In *PCR protocols: a guide to methods and applications* (ed. M. A. Innis, D. H. Gelfand, J. J. Sninsky, and T. J. White), p. 3. Academic Press, San Diego.
15. Orrego, C. (1990). In *PCR protocols: a guide to methods and applications* (ed. M.

A. Innis, D. H. Gelfand, J. J. Sninsky, and T. J. White), p. 451. Academic Press, San Diego.

16. Syvänen, A-C. and Söderlund, H. (1993). In *Methods in enzymology* (ed. R. Wu), Vol. 218, p. 474. Academic Press, New York.

17. Ihalainen, J., Siitari, H., Laine, S., Syvänen, A-C., and Palotie, A. (1994). *BioTechniques*, **16**, 938.

18. Syvänen, A-C., Tilgmann, C., Rinne, J., and Ulmanen, I. (1996). *Pharmacogenetics*, **7**, 65.

19. Syvänen, A-C., Söderlund, H., Laaksonen, E., Bengström, M., Turunen, M., and Palotie, A. (1992). *Int. J. Cancer*, **50**, 713.

20. Karttunen, L., Lönnqvist, L., Godfrey, M., Peltonen, L., and Syvänen, A-C. (1996). *Genome Res.*, **6**, 392.

21. Suomalainen, A., Majander, A., Pihko, H., Peltonen, L., and Syvänen, A-C. (1993). *Hum. Mol. Genet.*, **2**, 525.

22. Livak, K. J. and Hainer, J. W. (1994). *Hum. Mutat.*, **3**, 379.

23. Bertina, R. M., Koeleman, B. P. C., Koster, T., Rosendaal, F. R., Dirven, R. J., de Ronde, H., *et al.* (1994). *Nature*, **364**, 64.

24. Sitbon, G., Hustig, M., Palotie, A., Lönngren, J., and Syvänen, A-C. (1997). *Thromb. Haemost.*, **77**, 701.

25. Lagerkvist, A., Stewart, J., Lagerström-Fermer, M., and Landegren, U. (1994). *Proc. Natl. Acad. Sci. USA*, **91**, 2245.

26. Shumaker, J. M., Metspalu, A., and Caskey, C. T. (1996). *Hum. Mutat.*, **7**, 346.

27. Tully, G., Sullivan, K. M., Nixon, P., Stones, R. E., and Gill, P. (1996). *Genomics* **34**, 107.

28. Pease, A. C., Solas, E. J., Sullivan, M. T., Cronin, C. P., and Fodor, S. P. A. (1994). *Proc. Natl. Acad. Sci. USA*, **91**, 5022.

29. Southern, E. M., Maskos, U., and Elder, K. J. (1992). *Genomics*, **13**, 1008.

30. Pastinen, T., Kurg, A., Metspalu, A., Peltonen, L., and Syvänen, A-C. (1997). *Genome Res.*, **7**, 606.

Selective amplification of specific alleles

CYNTHIA D. K. BOTTEMA and STEVE S. SOMMER

1. Introduction

A variety of methods have been developed to detect known mutations and polymorphisms. Many of these methods are based on the polymerase chain reaction (PCR) because a segment of DNA can be amplified more than one millionfold for study or analysis. The amplified DNA may be analysed by chemical cleavage, restriction enzyme digestion, single-strand conformation polymorphism, gel electrophoresis, or hybridization to specific oligonucleotides.

These methods, however, all require additional steps following PCR. Therefore, one of the simplest and most rapid techniques is to modify the PCR for amplification of specific alleles by using specially designed oligonucleotides. This approach we have termed PCR amplification of specific alleles (PASA) (1, 2). Other laboratories also have developed this method to distinguish between specific alleles (3–6). Synonymous terms for PASA include allele-specific amplification (ASA), allele-specific PCR (ASP), and the amplification refractory mutation system (ARMS). The method is generally applicable for the detection of known point mutations, small deletions and insertions, polymorphisms, and other sequence variations.

1.1 Principle

For PASA, an oligonucleotide primer is designed to amplify preferentially one allele over another. Specificity is obtained if the oligonucleotide matches the desired allele, but mismatches the other allele(s) at or near the 3′ end of the allele-specific primer. The desired allele is readily amplified, while the other allele(s) is poorly amplified, if at all. The poor amplification is a result of the mismatch between the DNA template and the oligonucleotide. Since *Taq* DNA polymerase lacks any 3′→5′ exonuclease activity, the mismatch prevents efficient 3′ elongation by the *Taq* polymerase (*Figure 1*). Thus, single base changes, deletions, and insertions in DNA can be detected rapidly.

Using this technique, we have performed population screening, haplotype analysis, patient screening, and carrier testing for more than 65 different single

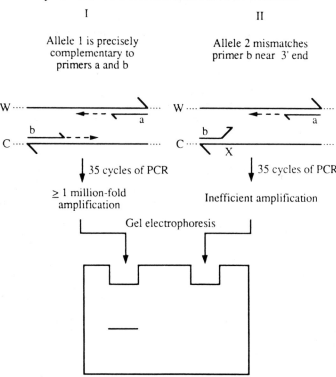

Figure 1. Schematic of polymerase chain reaction (PCR) amplification of specific alleles (PASA). The two antiparallel strands of chromosomal DNA are indicated by W, the Watson strand, and C, the Crick strand. The 5' to 3' direction is indicated by the half-arrows. Alleles 1 and 2 differ by a single base (represented by X where appropriate). Strands have been denatured by high temperature and the PCR oligonucleotide primers, a and b, have been annealed. In this diagram, elongation with *Taq* DNA polymerase, the third step in PCR, is underway. This process is represented by dashed lines that originate from the 3' end of the oligonucleotides. If the oligonucleotides are precisely complementary to the chromosomal DNA (panel I), elongation initiates efficiently. If there is a mismatch in the chromosomal DNA (X in panel II), elongation cannot be initiated efficiently for oligonucleotide b. After 35 cycles of PCR, a one to ten million fold amplification occurs in panel I, where as much less amplification occurs in panel II. When an aliquot of the PCR is electrophoresed, an abundance of the amplified DNA segment of appropriate size can be detected in panel I by staining with ethidium bromide. In contrast, no segment is detected in panel II. Thus, a mutation (or polymorphism) can be detected by using an appropriately designed oligonucleotide that promotes efficient elongation from a mutant chromosome and inefficient elongation from the normal chromosome. From ref. 1 with permission.

base alleles (7). Allele-specific conditions were found for each of the assays attempted. The technique can be further adapted to obtain haplotypes without samples from family members, to detect two alleles in one reaction tube, to distinguish single base changes at more than one site, and to use a number of PCR product separation and/or detection systems.

Table 1. Oligonucleotides specific for the transition at amino acid 397 of the factor IX gene

Primer[a]	Full oligonucleotide name[a]	Sequence[b]
A: $17(G^n)$	F9(Hs)-E8(31 327)-17U(G^n)	GGA TAC CTT GGT ATA T<u>G</u>
B: $17(G^{n-1})$	F9(Hs)-E8(31 328)-17U(G^{n-1})	GA TAC CTT GGT ATA T<u>G</u>T
C: $17(G^{n-2})$	F9(Hs)-E8(31 329)-17U(G^{n-2})	A TAC CTT GGT ATA T<u>G</u>T T
D: $17(G^{n-3})$	F9(Hs)-E8(31 330)-17U(G^{n-3})	TAC CTT GGT ATA T<u>G</u>T TC
E: $18(A^n)$	F9(Hs)-E8(31 326)-18U(A^n)	G GGA TAC CTT GGT ATA T<u>A</u>

[a] The trivial names are used in the text for ease of reading. The corresponding informative full name allows determination of the sequence of the oligonucleotide, its allele specificity, its orientation (downstream or upstream relative to transcription), and the size of amplified segment expected from a particular amplification reaction (26). The nomenclature is of the form: G(O)-R(C)-SD(M) where G = gene abbreviation, O = organism, R = region in gene {5', exon (E), intron (I), or 3'], C = location of the 5' complementary base [base number from gene sequence], S = total length of oligonucleotide [number of bases in primer], D = 5' to 3' direction of oligonucleotide [downstream or sense direction (D), upstream or antisense direction (U)], and M = base and point of mismatch from the 3' end [penultimate base (n)]. If non-complementary bases are present (as in the PAMSA primers), then (I-L) may be added to the full oligonucleotide name where I-L = identifier and length of non-complementary bases. As an example of the full oligonucleotide names, A = F9(Hs)-E8(31 327)-17U(G^n). As the full name indicates, the primer is in exon 8 of the human factor IX gene and begins at base pair 31 327. The primer is 17 bases long and the sequence is in the upstream (antisense) direction. The primer's specificity is imparted by the G at the 3' end.
[b] The G that is underlined matches the T^{397} mutant allele and mismatches the normal I^{397} allele. The A that is underlined in $18(A^n)$ matches the normal allele and mismatches the mutant allele. Adapted from ref. 2 with permission.)

2. Parameters of PASA

It was hypothesized that a mismatch at or near the 3' end of the oligo-nucleotide would be likely to hinder the 3' elongation by *Taq* polymerase. In order to establish the critical parameters that determine the specificity of this interaction, a T→C transition in the X-linked factor IX gene at nucleotide 31 311 was chosen as a test system. The T→C transition substitutes threonine for isoleucine at amino acid 397 and causes haemophilia B in males with this mutation. Two primers were synthesized to be identical to the antisense strand for the normal and mutant alleles by differing at the 3' base (*Table 1*). Primer 18 (A^n) is specific for the normal I^{397} allele in the factor IX gene, whereas primer 17 (G^n) is specific for the mutant T^{397} allele. If primer 17 (A^n) is used, specific amplification occurs only in male patients with the mutation. In females carrying this mutation, amplification occurs with both primers because they are heterozygous at this site (*Figure 2*).

2.1 Point of mismatch and oligonucleotide length

To examine whether single base mismatches could reproducibly and dramatic-ally interfere with DNA amplification by PCR, additional oligonucleotides were synthesized with mismatches to the target DNA located at different

Cynthia D. K. Bottema and Steve S. Sommer

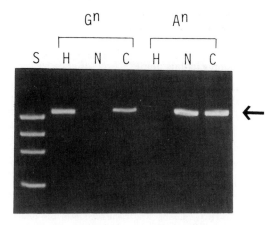

Figure 2. Carrier testing for haemophilia B by PASA. PASA was used to detect a T→C transition in the X-linked factor IX gene at nucleotide 31 311 which substitutes threonine for isoleucine[397] and causes haemophilia B in males. PCR was performed with either the primer 17 (G^n) or the primer 18 (A^n) as indicated. Primer 17 (G^n) amplifies the mutant 5^{397} allele and primer 18 (A^n) amplifies the normal I^{397} allele. Lane S: standard, 250 ng φX174 *Hae*III restriction fragments. Lane H: haemophiliac with the mutant T^{397} allele. Lane N: non-carrier granddaughter of the haemophiliac. Lane C: carrier granddaughter of the haemophiliac. The arrow indicates the expected size of the amplified DNA segment. From ref. 27 with permission.

positions within the primers (2). The oligonucleotides were specific for the mutant 'C' at position 31 311 in factor IX gene (T^{397} allele), but the point of mismatch with the normal allele was not at the 3' base (*Table 1*).

To examine the effect of oligonucleotide length on specificity, the second target chosen was an A→T transversion in the factor IX gene at nucleotide 48. This transversion does not alter an amino acid and was found as a sequence variant in a normal male. PCR primers were synthesized to be identical to the antisense strand for the variant and normal alleles by containing either a 'T' or an 'A' to match or mismatch the target DNA at nucleotide 48 (*Table 2*). The primers were designed to vary not only in the point of mismatch, but also in length.

The following conclusions were drawn from these studies (*Table 2*) (1, 2):

(a) When the 3' or the 3' penultimate base of the oligonucleotide mismatched a target allele, no amplification product could be detected.

(b) When the mismatches were three or four bases from the 3' end of the primer, differential amplification still was observed, but only at certain concentrations of magnesium chloride.

(c) A primer as short as 13 nucleotides ($T_m = 36\,^{\circ}C$) was effective, although high magnesium concentrations were required (> 4.5 mM).

164

Table 2. Comparison of oligonucleotide length and the point of mismatch

Oligonucleotide[a]	Factor IX allele[b]	$[Mg^{+2}]$ window (mM)[c]	Reference T_m (°C)[d]
17 (G^n)	bp 31311: C	> 2.5	48
17 (G^{n-1})	bp 31311: C	> 3.0	46
17 (G^{n-2})	bp 31311: C	> 3.0	44
17 (G^{n-3})	bp 31311: C	1.0	46
18 (A^n)	bp 31311: T	> 2.0	50
15 (A^n)	bp 48: T	> 2.0	42
15 (A^{n-1})	bp 48: T	> 2.0	44
15 ($A^{n-2, n-10, n-11}$)	bp 48: T	1.0	42
15 (A^{n-4})	bp 48: T	0.7	44
15 (A^{n-7})	bp 48: T	0.2	44
14 (A^n)	bp 48: T	> 2.0	40
13 (A^n)	bp 48: T	> 2.0	36
12 (A^n)	bp 48: T	–	34
15 (T^n)	bp 48: A	> 2.0	42
15 (T^{n-1})	bp 48: A	> 2.0	44

[a] The number indicates the length of the oligonucleotide; the letter indicates the mismatched base; 'n' indicates the point of the mismatch relative to the 3' end of the primer.
[b] The T→C transition at base 31311 in the factor IX gene substitutes threonine for isoleucine 397. An A→T transversion at base 48 in the factor IX gene is a rare sequence variant found in normal individuals.
[c] $[Mg^{2+}]$ window is the difference in minimum Mg^{2+} concentrations required for amplification of the specific allele versus the mismatched allele.
[c] Melting temperatures were estimated under standard conditions (i.e. 1 M NaCl) from 4°C × (G + C) + 2°C × (A + T).
Adapted from ref. 10 with permission.

2.2 Detection limit

To determine if genomic samples from multiple individuals could be simultaneously analysed for screening populations, 250 ng of normal genomic DNA was mixed with decreasing concentrations of genomic DNA from an individual hemizygous for the factor IX sequence variant at position 48 (2). With as little as 6.25 ng of genomic DNA with the sequence variation, a specific product could be seen in an agarose gel stained with ethidium bromide (*Figure 3*). Thus, the matched allele can be detected in the presence of a 40-fold excess of the mismatched allele.

With ethidium bromide staining, we typically pool samples from four individuals, thereby conserving substantial residual sensitivity to compensate for interindividual variation in DNA yields. In this manner, 1000 individuals can be screened per day by manual methods (8). However, the sensitivity can be increased in a number of other ways, including radiolabelling or staining with fluorescent dyes other than ethidium bromide.

To ascertain whether high concentrations of target DNA could overcome the specificity of PASA, a normal cDNA clone of factor IX was serially diluted by factors of ten (2). Using the mutant primer 17 (G^n) for the T^{397} site,

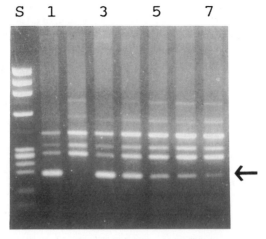

Figure 3. Effect of allele concentration on PASA. A polymorphic T→A transversion occurs in the X-linked factor IX gene at base 48. The T allele is common in the population, while the A allele is a rare variant. To determine if multiple genomic samples could be simultaneously analysed for population screening of the rare A allele, an oligonucleotide specifically matching the A allele, primer 15 (A^{n-1}), was designed. PASA was performed with the primer 15 (A^{n-1}) using decreasing concentrations of DNA from individual A with the rare A allele in the presence of 250 ng DNA from individual T with the common T allele. Lane S: standard, 250 ng ϕX174 *Hae*III restriction fragments. Lane 1: DNA from individual A with the rare A allele (250 ng). Lane 2: DNA from individual T with the common T allele (250 ng). Lane 3: DNA from individual A and from individual T (250 ng/250 ng). Lane 4; A/T (83 ng/250 ng). Lane 5: A/T 925 ng/250 ng). Lane 6: A/T (12.5 ng/250 ng). Lane 7: A/T (6.125 ng/250 ng). The arrow indicates the size of the expected amplified segment. Other bands are spurious. Results indicate that even with as little as 6.25 ng DNA from individual A, a specific product could be seen. From ref. 2 with permission.

an amplified product of appropriate size from normal DNA could be detected when the concentration of target DNA was 10 000-fold or higher than the minimum concentration of DNA required for amplification with the normal matched primer 18 (A^{n}). At other sites, however, lower concentrations of DNA were able to overcome specificity (see Section 5.1.1).

3. PASA protocol

PASA is performed as a standard polymerase chain reaction (PCR) with slight modifications to guarantee sensitivity. Two factors are essential:

(a) The DNA must be of high quality to ensure a successful reaction if the matching primer is present.

(b) The oligonucleotides must be appropriately designed to ensure that the reaction is specific.

3.1 DNA extraction

The DNA extraction protocol is important for obtaining reproducible PCR amplification for each sample. We routinely use blood lymphocytes as the source of DNA. However, it should be noted that the type of anticoagulant used in blood collection is critical for the quality of the DNA (see *Protocol 1*).

Protocol 1. DNA extraction from blood cells for PASA

Equipment and reagents

- Vacutainer tubes containing acid citrate dextrose (ACD) solution B or solution A: 0.48% citric acid, 1.32% sodium citrate, 1.47% glucose (Becton Dickinson)
- PBS: 1 mM KH_2PO_4, 154 mM NaCl, 5.6 mM Na_2HPO_4 pH 7.4
- Pronase: dissolve 20 mg/ml pronase (Boehringer Mannheim) in distilled water without pre-digestion, and store in small aliquots at –20°C until used[a]
- TE buffer: 10 mM Tris–HCl, 1 mM EDTA pH 8.0

- RNase A: dissolve 10 mg/ml RNase A (Sigma) in 10 mM Tris–HCl, 15 mM NaCl pH 7.5. Heat the dissolved RNase A to 100°C for 15 min, allow to cool slowly to room temperature, dispense into aliquots, and store at –20°C until used.
- Phenol: saturate phenol (Gibco BRL) with TE buffer pH 8, add 0.2% β-mercapto-ethanol, and 0.05% β-hydroxyquinolone
- Chloroform: chloroform:isoamyl alcohol (24:1)

Method

1. Draw blood in 8 ml or 10 ml vacutainer tubes containing acid citrate dextrose (ACD) solution B or solution A. The anticoagulant should be thoroughly mixed with the blood by inverting the tube several times. This anticoagulant enhances the stability of DNA in blood (9).

2. The tubes may remain at 4°C or room temperature for up to a week. Alternatively, the tubes may be frozen at –20°C to –70°C indefinitely.

3. Thaw the blood. Transfer the blood to a sterile polypropylene tube. Dilute the blood with 1 vol. of PBS. Mix the solution by inverting tube.

4. Centrifuge the tube at 3400 *g* for 15 min at room temperature.

5. Carefully pour off the supernatant which contains lysed erythrocytes. Resuspend the leucocyte pellet in 0.6 times the original blood volume of TE buffer.

6. Disperse all clumps by drawing the solution into a pipette several times. Add pronase to the tube at a final concentration of 500 μg/ml. If residual mRNA is a concern, add RNase A to the tube at a final concentration of 50 μg/ml. Mix the solution by inverting the tube several times.

7. Add 10% SDS to the tube at a final concentration of 0.5%, and mix well by inverting the tube several times. The contents of tube will become viscous.

8. Incubate the tube at least 2 h in a 37°C water-bath.

Protocol 1. *Continued*

9. Add an equal volume of phenol to the sample, and shake the tube by hand at approximately 150 cycles per minute for 3 min. This rate strikes a balance between thorough, rapid deproteination and shearing of DNA. If it is necessary to obtain DNA whose average size is more than 40 kb, shake more gently for a longer period of time.

10. Centrifuge the tube at 3400 *g* for 10 min at room temperature. Draw off the upper aqueous phase and transfer to another polypropylene tube.

11. Repeat the phenol extraction. Draw off the upper aqueous phase and transfer to another polypropylene tube.

12. Add an equal volume of chloroform to the aqueous phase. Shake the tube by hand for 3 min and centrifuge the tube at 3400 *g* for 5 min at room temperature.

13. Draw off the upper aqueous phase and transfer to another polypropylene tube. Determine the volume of the sample and adjust the sodium chloride concentration to 0.1 M.

14. Add 2 vol. of ice-cold (−20 °C) 95% ethanol to the tube, and mix the contents well by inverting the tube.

15. Incubate the tube at least 30 min at −20 °C until a visible precipitate appears.

16. Centrifuge the tube at 3400 *g* for 10 min at 4 °C. Decant the supernatant, and drain the pellet carefully.

17. Rinse the pellet with ice-cold 70% ethanol and lyophilize.

18. Resuspend the pellet with a pipette in one-tenth of the original blood volume of 10 mM TE buffer until no visible pellet remains. To assure that the DNA has completely dissolved, rotate the sample for at least 4 h at room temperature. For human blood, the average yield will be approximately 30 μg DNA/ml of blood.

[a] Pre-digestion of the pronase removes the desired RNase activity.

3.2 PCR amplification of specific alleles

For specificity, the design of the PASA oligonucleotides is crucial (see Section 2.1). Normally, the oligonucleotide is designed such that the mismatch is at the 3′ or 3′ penultimate base. The oligonucleotide is also usually greater than 14 bases in length.

For each pair of PCR primers, the allele-specific PCR oligonucleotide is generally designed to have an estimated melting temperature (T_m) in the range of 44–48 °C under standard conditions (1 M sodium chloride). The T_m is calculated as $4 °C (G + C) + 2 °C (A + T)$. The other primer that does not anneal at the polymorphic site is designed to have a melting temperature of 48 °C.

It is equally important to consider the concentration of the source DNA. When initially optimizing the conditions of PASA, a final concentration of 10 ng/μl is used (see *Protocol 2*). However, if attempting population screening (see Section 2.2) or trying to increase sensitivity (see Section 3.3), the DNA concentration may need to be adjusted.

Protocol 2. PCR amplification of specific alleles (PASA)

Equipment and reagents

- AmpliTaq DNA polymerase (Perkin Elmer Cetus)
- Genomic DNA (see *Protocol 1*)
- Loading dye: 10% glycerol, 0.025% bromophenol blue, 0.025% xylene cyanol FF
- TAE buffer: 40 mM Tris–acetate, 1 mM EDTA pH 8.0
- Ethidium bromide: stock solution of 10 mg/ml dH$_2$O

Method

1. Add genomic DNA (250 ng) to a 500 μl microcentrifuge tube in a 25 μl total volume of 50 mM potassium chloride, 10 mM Tris–HCl pH 8.3, 1.5–3.5 mM magnesium chloride, 200 μM of each deoxyribonucleotide (dATP, dTTP, dGTP, and dCTP), 0.05–1.00 μM of each primer, and 0.50 U of AmpliTaq DNA polymerase.

2. Mix the reaction by briefly vortexing the tubes and centrifuging in a microcentrifuge for a few seconds.

3. Gently add two drops of mineral oil to the tube, overlaying the contents. The mineral oil reduces evaporation of the contents during the heating.

4. Perform 35 cycles of 1 min denaturation at 94°C, 2 min annealing at 50°C, and 3 min elongation at 72°C in an automated thermal cycler, with one final 10 min elongation at 72°C.[a]

5. Upon completion of the cycles, place 5–10 μl of each reaction in a fresh microcentrifuge tube. Care must be taken to avoid any carry-over of the mineral oil.

6. Add 1 μl of the loading dye to the tube. Vortex the contents briefly and centrifuge for a few seconds in a microcentrifuge.

7. Load the sample and appropriate DNA size markers into separate wells of a 1–3% (w/v) agarose gel and electrophorese in TAE buffer.

8. Visualize the amplification products by staining the agarose gel with ethidium bromide (0.5 μg/ml dH$_2$O or TAE buffer) for 30 min at room temperature and examining the gel under UV light. Optimization and troubleshooting are discussed below (see Section 3.3).

[a] 30–40 cycles of PCR are performed with little observable difference. Consequently, 35 cycles of PCR can be used routinely.

The standard primer concentration is 1 μM for most PCR protocols. However, it was observed that decreasing the total oligonucleotide concentration to 0.05–0.25 μM could increase specificity. Consequently, the concentration should be determined experimentally for each primer pair. Generally, 0.10 μM of each primer is sufficient to obtain detectable and specific amplification.

Likewise, the stringency of the amplification can be altered by the final magnesium concentration. Therefore, the magnesium concentration should be optimized for each primer pair. Frequently, a magnesium concentration of 2.5 mM will ensure that specific amplification will occur if the matching primer is present.

3.3 Optimization of PASA

In the majority of cases, a standard magnesium titration (1.5, 2.5, 3.5, and 4.5 mM) and oligonucleotide titration (1.00, 0.25, 0.10, and 0.05 μM) are

---

Figure 4. Optimization of PASA. A mutation at base pair 6461 (G→A transition) in the X-linked factor IX gene substitutes a glutamine for arginine[29], causing haemophilia B. An oligonucleotide was designed to specifically match the mutant A allele (primer II). Two non-specific oligonucleotides were also designed (primers I and III). Using these primers, the conditions for PASA specificity were tested. S: standard, 250 ng φX174 HaeIII restriction fragments. A: DNA from individual with the mutant A allele. G: DNA from individual with the normal G allele. The arrows indicate the expected size of the specific fragments. (A) Effect of oligonucleotide concentration. MgCl$_2$ concentration was 3.5 mM. Primers I and II were used. Lanes 1–4: mutant A allele. Lanes 5–8: normal G allele. Lanes 1 and 5, 1.0 μM primers; lanes 2 and 6, 0.25 μM primers; lanes 3 and 7, 0.10 μM primers; lanes 4 and 8, 0.05 μM primers. (B) Effect of MgCl$_2$ concentration. Oligonucleotide concentrations of I and II were 0.25 μM. Lanes 1–4: mutant A allele. Lanes 5–8: normal G allele. Lanes 1 and 5, 1.5 mM MgCl$_2$; lanes 2 and 6, 2.5 mM MgCl$_2$; lanes 3 and 7, 3.5 mM MgCl$_2$; lanes 4 and 8, 4.5 mM MgCl$_2$. (C) Effect of non-specific oligonucleotide choice. Primer I was replaced by another non-specific oligonucleotide, primer III. Primers II and III amplify a segment 21 bases longer than primers I and II (as indicated by the arrow). MgCl$_2$ concentration was 3.5 mM. Primers I and II were used. Lanes 1–4: mutant A allele. Lanes 5–8: normal G allele. Lanes 1 and 5, 1.0 μM primers II and III; lanes 2 and 6, 0.25 μM primers II and III; lanes 3 and 7, 0.10 μM primers II and III; lanes 4 and 8, 0.05 μM primers II and III. (D) Effects of DNA, Taq polymerase, and deoxynucleotide concentrations. Primers I and II concentrations were 0.25 μM and MgCl$_2$ concentration was 3.5 mM. DNA, Taq polymerase, and dNTPs concentration were varied separately. Lanes 1–7: mutant A allele. Lanes 8–14: normal G allele. Lanes 1 and 8, no change; lanes 2 and 9, 25 ng DNA (10% of normal concentration); lanes 3 and 10, 12.5 ng DNA; lanes 4 and 11, 0.25 U Taq polymerase (50% of normal); lanes 5 and 12, 1.0 U Taq polymerase; lanes 6 and 12, 0.625 mM dNTPs (50% of normal); lanes 7 and 14, 0.25 mM dNTPs. (E) Effect of the addition of formamide or a second pair of unrelated oligonucleotides. Primers I and II concentrations were 0.25 μM and MgCl$_2$ concentration was 3.5 mM. Either formamide or a second pair of unrelated, non-specific oligonucleotides was included in the PCR. The second pair of primers amplified a 313 bp non-polymorphic segment in intron 5 of the factor IX gene. Lanes 1–4: mutant A allele. Lanes 5–8: normal G allele. Lanes 1 and 5, no additions; lanes 2 and 6, 2% formamide; lanes 3 and 7, 4% formamide; lanes 4 and 8, 0.1 μM second primer pair. From ref. 7 with permission.

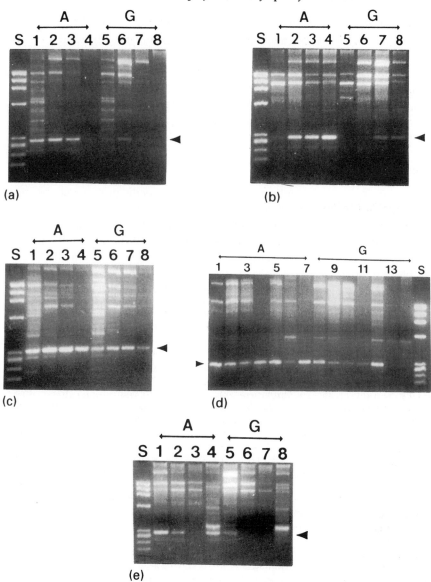

(a)

(b)

(c)

(d)

(e)

sufficient to define conditions for both a robust and specific amplification (*Figure 4A* and *B*). Ideally, magnesium and oligonucleotide concentrations are adjusted also to produce spurious amplified fragments that do not interfere with allele detection, but serve as an internal control for the technical success of the PCR when the specific band is absent.

Since the cost of oligonucleotides is low compared to labour and time costs, we typically order PASA oligonucleotides in both directions to maximize the likelihood that only a magnesium and oligonucleotide titration is necessary to optimize the assay. However, if further optimization is required to obtain specificity (7, 10), a variety of parameters can be modified, including:

(a) Allele-specific oligonucleotide. Designing the allele-specific primer is critical. Lowering the T_m from 48°C to 42–44°C and placing the mismatch at the 3′ base will increase specificity. Occasionally, a segment does not amplify well. Designing new primers using the other strand for the allele-specific mismatch may provide better amplification.

(b) Non-specific oligonucleotide. Occasionally, a given pair of primers will not be specific. Surprisingly, replacement of the non-specific primer with another oligonucleotide at a new location will often provide specificity (*Figure 4C*). Again replacement of the non-specific primer may be more cost-effective than further optimization. Generally, primer pairs are chosen to yield products that range in size between 300 and 600 bases. However, this is not always possible as in the case of double PASA (see Section 4.2.3). We have specifically amplified segments from 200 to 2700 bases in length.

(c) Magnesium concentration. Specificity can sometimes be achieved by lowering the magnesium concentration below 1.5 mM. Adding EDTA is a simple way of decreasing the 'effective' magnesium concentration without making a different PCR buffer/salt stock solution. Likewise, increasing the magnesium concentration above 4.5 mM can produce non-interfering spurious amplification products which are useful internal controls.

(d) Oligonucleotide concentration. Decreasing the oligonucleotide concentration to 0.05 μM may increase specificity. However, below 0.025 μM the amplification signal generally becomes weak or undetectable by ethidium bromide staining.

(e) DNA concentration. A tenfold dilution of the standard genomic DNA concentration (i.e. 1 ng/μl) can increase specificity and still provide an adequate amplification signal (*Figure 4D*). Diluting the template can also avoid problems caused by contamination of the DNA with any PCR inhibitors.

(f) *Taq* polymerase. Decreasing the amount of enzyme in each reaction (0.2–0.3 U/μl) can increase specificity, while adding more enzyme can create spurious bands to serve as internal controls (*Figure 4D*).

(g) Deoxynucleotide concentration. Decreasing the concentrations of the dNTPs to 25–50 μM can prevent spurious amplification (*Figure 4D*).

(h) Formamide or DMSO. Inclusion of formamide (1–5%) or DMSO (2–10%) can increase the signal strength and eliminate undesired spurious bands, especially if the G + C content is high. In addition, allele specificity may be enhanced (*Figure 4E*).

(i) Additional pair of primers. A second pair of compatible non-specific primers can be added to generate a constant band. This serves both as an internal control for the technical success of the PCR and to in-crease specificity by providing another substrate for the *Taq* polymerase (*Figure 4E*).

(j) Addition of a competing oligonucleotide. Specificity can be enhanced by including both allele-specific oligonucleotides in the same PCR (11, 12). This approach is called competitive oligonucleotide priming (COP). The allele-specific oligonucleotides compete for their respective templates, reducing the opportunity for false priming events. The competition can increase specificity, particularly if specificity is being overwhelmed due to relatively low concentration of target DNA (11). The allele-specific PCR products can be distinguished either by differentially labelling the allele-specific oligonucleotides or designing the allele-specific oligonucleotides to be different lengths (see Section 5.1.1 on PCR amplification of multiple specific alleles, PAMSA).

(k) Nested PASA. Using a PCR product as the source of DNA template for a nested PASA can provide specificity, particularly if the region to be amplified is highly repetitive. Note, however, that the concentration of DNA is critical and a 10^6-fold or greater dilution may be required of the original PCR product.

(l) Number of amplification cycles. Decreasing the number of PCR cycles may reduce detection of any minor amplification of the mismatched allele. Often, however, the number of cycles makes little difference in specificity.

(m) Annealing temperature. Raising the PCR annealing temperature can increase specificity. This may be undesirable if these amplification reactions are incompatible with reactions optimized at different cycle parameters. In our experience with more than 100 PASA reactions, it has never been necessary to deviate from our standard cycles of 1 min at 94 °C, 2 min at 50 °C, and 3 min at 72 °C, except to increase elongation times for segments greater than 1.5 kb or segments with a G + C content of 60%.

(n) PCR enhancing reagents. Inclusion of a PCR enhancing reagent has been shown to increase specificity and eliminate spurious bands during PASA (13).

In general, optimization of any of a number of the above parameters can achieve the desired results of adequate amplification and detection of specific alleles. In our laboratory, conditions for allele specificity were found readily for all the sites we have analysed.

4. Implementation of PASA

We have successfully used PASA at 41 sites to perform:

(a) Haplotyping in the transthyretin gene (ten sites), the factor IX gene (two sites), the dopamine D_2 receptor gene (three sites), and the factor VIII gene (one site).

(b) Population screening for rare variants in the factor IX gene (four sites).

(c) Carrier testing and population screening for mutations in the phenylala-nine hydroxylase gene (six sites), the transthyretin gene (six sites), the factor IX gene (seven sites), and the factor VIII gene (two sites) (7).

4.1 Generality of PASA

One or two alleles were assayed at each of the above mentioned sites for a total of 69 allele-specific assays (*Table 3*) (7). Reproducible discrimination between all possible single base changes has been obtained using PASA from a total of 25 transversion alleles and 44 transition alleles when the 3' or 3' penultimate base of the oligonucleotide primer matched the desired allele.

All combinations of mismatched primer and template pairs were successful (*Table 3*). Because mutations and polymorphisms are commonly due to tran-sitions at CpG dinucleotides, A:C, G:T, C:A, and T:G primer:template mis-matches were common in our sample. The results differ from those of Kwok *et al.* (14) who made all the mismatched combinations at one site of an HIV-1 sequence. They found that differential amplification of the perfect match occurred only when the mismatch was A:G, G:A, C:C, and A:A. The reasons for the discrepancy are unclear. Optimization of DNA concentrations and dNTP concentrations were reported by Kwok *et al.* (14). However, the oligo-nucleotide concentrations apparently were not examined and were higher than is usually optimal (0.5 μM versus 0.1 μM). It is possible that lowering the oligonucleotide concentration and other optimization measures are critical for obtaining specificity at their HIV-1 site.

Table 3. Generality of PASA.[a]

PASA primer[b]	Mismatched template allele[c]				Number of transitions[d]	Number of transversions[d]
	T	A	C	G		
A	–	1	15	4	15	5
T	1	–	4	11	11	5
G	9	4	–	4	9	8
C	2	9	5	–	9	7
				Total	44	25

[a]Number of PASA reactions where specificity obtained for the corresponding primer:template mis-match.
[b]Sequence of the allele-specific PASA oligonucleotide.
[c]Sequence of the mismatched DNA template.
[d]Relationship of the two template alleles being distinguished. For example, distinguishing C and T transition alleles in genomic DNA would involve mismatches between an 'A' primer and the 'C' allele template and between a 'G' primer and the 'T' allele template. That is, specificity is obtained for A:C and G:T primer:template mismatches.
Adapted from ref. 7 with permission.

4.2 Applications

PASA has been applied to a wide variety of diagnostic and experimental conditions. When a single mutation or polymorphism is to be detected in a relatively few samples, the implementation of PASA is straightforward. However, if even more than one mutation or polymorphism is to be examined and/or many samples are to be screened, PASA can readily accommodate for these requirements. In the examples given below, PASA allowed the simultaneous testing of more than one site, screening of multiple chromosomes, and haplotyping of single chromosomes.

4.2.1 Individual screening by PASA

The ease and technical simplicity of PASA allow rapid patient screening and carrier testing for known mutations or polymorphisms. If there is more than one mutation in a family, the mutations can be analysed separately. However, in cases where mutations are clustered within a gene, it is most efficient to detect the mutations simultaneously. For example, two mutations ($Arg^{408} \rightarrow Trp^{408}$ and an intron 12 splice junction defect) in the phenylalanine hydroxylase gene which cause phenylketonuria (PKU) are within 1.4 kb of each other. By designing the allele-specific and non-specific oligonucleotides appropriately, it was possible to screen family members for these two mutations simultaneously (*Figure 5*) (1).

4.2.2 Population screening by PASA

PASA is also amenable to population screening since the presence of a mutant allele can be detected even in high concentrations of normal DNA. For instance, the Trp^{408} mutation and intron 12 splice junction defect in the phenylalanine hydroxylase gene can be detected in the presence of a 40-fold excess of the normal gene. This allows efficient population screening for these two mutations and detection of PKU carriers (*Figure 6*) (1).

It is possible to screen over 400 chromosomes in 50 tubes by amplifying four individuals per PCR. Thus, one person was able to screen 800 transthyretin alleles in one day for mutations associated with familial amyloidotic polyneuropathy (8).

4.2.3 Haplotyping by double PASA

Without DNA samples from appropriate family members, determining the linkage of haplotypes can be difficult and laborious. PASA can be adapted to provide the haplotype of an individual in the absence of relatives by utilizing pairs of allele-specific primers (15, 16). This 'double PASA' differentially amplifies each haplotype (*Figure 7*). Four amplifications can distinguish the haplotypes produced by a pair of biallelic polymorphisms.

Haplotypes are selectively amplified with PCR primers specific for the relevant alleles. For example in *Figure 7*, haplotype I can be detected with primers

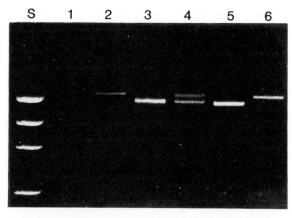

Figure 5. Simultaneous detection of two phenylketonuria (PKU) mutations by PASA. Two mutations, Arg[408]→Trp[408] and g→a in the intron 12 splice junction, were found in the phenylalanine hydroxylase gene of a PKU patient. Family members were simultaneously screened for these two mutations using three oligonucleotides. Oligonucleotides I and II were specific for the Trp[408] and intron 12 splice junction mutations, respectively. Oligonucleotide III was the non-specific oligonucleotide for both reactions. When PCR was performed with oligonucleotides I and III, a 1.4 kb segment was detected if the Trp[408] mutant allele was present. When PCR was performed with oligonucleotides II and III, a 1.3 kb segment was detected if the intron 12 splice junction mutant allele was present. Oligonucleotide concentration was 0.10 μM each. Lane S: standard, 250 ng φX174 HaeIII restriction fragments. Lane 1: DNA from unrelated non-carrier control; oligonucleotides I, II, and III. Lane 2: DNA from PKU patient with two mutations; oligonucleotides I and III. Lane 3: DNA from PKU patient with two mutations; oligonucleotides II and III. Lane 4: DNA from PKU patient with two mutations; oligonucleotides I, II, and III. Lane 5: DNA from family member carrying intron 12 splice junction mutation; oligonucleotides I, II, and III. Lane 6: DNA from family member carrying Trp[408] mutation; oligonucleotides I, II, and III. From ref. 1 with permission.

1 and 3; haplotype II can be detected with primers 2 and 3; haplotype III can be detected with primers 1 and 4; and haplotype IV can be detected with primers 2 and 4. Using two polymorphisms (PM1 and PM2) in the human dopamine D_2 receptor, six subjects were chosen for analysis (*Figure 8*). The haplotypes and primers were as indicated in *Figure 7*. DNA from these individuals was amplified with each of the four sets of primers (one set per gel quadrant in *Figure 8*). Amplified segments were seen in individuals 4, 5, and 6 in the upper left quadrant of the gel. These individuals had haplotype I. Likewise, amplified segments in the upper right quadrant of the gel were seen for individuals 1, 2, and 4. These individuals had haplotype II. From the lower left quadrant, none of the individuals had haplotype III. From the lower right quadrant, individuals 2, 3, and 6 had haplotype IV.

Double PASA is an important tool for haplotyping doubly heterozygous individuals because the physical linkage of alleles on a strand of DNA is necessary to determine the haplotype. In the example above (*Figures 7* and *8*),

(A) I & III   (B) II & III   (C) II & III

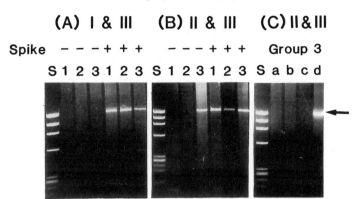

Figure 6. Population screening for carriers of two phenylketonuria (PKU) mutations by PASA. DNA from unrelated normal subjects was screened for two mutations found in the phenylalanine hydroxylase gene of a PKU patient (Arg^{408}→Trp^{408} and g →a in the intron 12 splice junction) using oligonucleotides I, II, and II as described in *Figure 5*. PASA was performed on groups of four individuals using 60 ng of DNA each (lanes indicated spike–). As a positive control, PASA was repeated with each group after the addition of 60 ng of DNA from the PKU patient (lanes indicated spike +). Lane S: standard, 250 ng φX174 *Hae*III restriction fragments. Lane 1: group 1. Lane 2: group 2. Lane 3: groups 3. (A) Screening for the Trp^{408} mutant allele by PASA. PCR was performed with oligonucleotides I and III (0.25 μM each). No carriers were detected. (B) Screening for the intron 12 splice junction mutant allele by PASA. PCR was performed with oligonucleotides II and III (0.05 μM each). A carrier of this mutation was detected in group 3. (C) Identifying the PKU carrier in group 3. Using oligonucleotides II and III, 250 ng of DNA from each person in group 3 (a–d) was screened by PASA. Arrow indicates size of expected amplified segment. From ref. 1 with permission.

physical linkage is required to distinguish doubly heterozygous individuals with haplotypes I and IV (such as individual 6) from those with haplotypes II and III.

We have demonstrated double PASA with two polymorphisms in the human dopamine D$_2$ receptor gene (15) and six polymorphisms in the transthyretin gene (8). The six polymorphisms in the transthyretin gene were analysed to determine the haplotype of a rare transthyretin gene variant, TTR M^{119}. For the six polymorphisms in the transthyretin gene, it was necessary to perform 24 combinations of double PASA (*Figure 9*). The results confirmed that all the TTR M^{119} alleles had the same haplotype (T-G-G-A-C-G), suggesting a common ancestor for this rare gene variant (*Table 4*) (8).

Double PASA should be generally applicable for haplotyping, provided that the segment between the polymorphisms can be amplified at least to a modest extent. If the polymorphic sites are separated by too great a distance to allow PCR amplification, double PASA can be combined with inverse PCR (16). By circularization, the genomic targets can be placed close enough together to allow amplification.

A

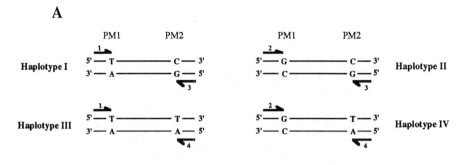

B

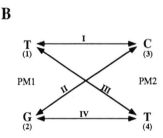

Figure 7. Schematic of double PASA. (A) In this example, there are two biallelic polymorphic sites. T and G are present on the sense strand at polymorphic site 1 (PM1). C and T are present at polymorphic site 2 (PM2). PCR primers 1–4 are synthesized. The half-arrowhead signifies the 3′ end of the oligonucleotides. Primers 1 and 2 are identical, except for the T and G at their 3′ ends, respectively. Likewise, primers 3 and 4 are identical, except for the G and A at their 3′ ends, respectively. (B) The possible haplotypes (I–IV) can be differentially amplified by four PCRs with the indicated primers. From ref. 15 with permission.

5. Modifications of PASA

PASA is a generally applicable method for detecting point mutations, rare variants, and polymorphisms. However, PASA does have certain disadvantages:

(a) If the specific allele is absent and spurious bands of other sizes are not produced, another set of compatible primers must be added to generate a constant band to serve as an internal control for the technical success of the PCR.

(b) Two PCR reactions are required to determine if an individual is heterozygous or homozygous for one of the alleles.

(c) The specificity of PASA can be overwhelmed by high concentrations of template DNA. If the template DNA concentrations vary substantially, this can be a major problem.

S N 1 2 3 4 5 6 S N 7 8 9 10 11 12 S

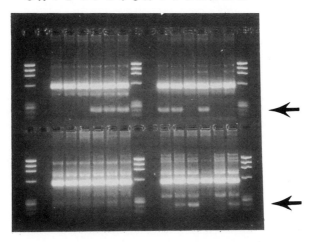

S N 13 14 15 16 17 18 S N 19 20 21 22 23 24 S

Figure 8. Haplotyping with double PASA. The gel is divided into four quadrants. In each quadrant, a different set of PCR primers detect one of the four possible haplotypes (see *Figure 7*). The *arrow* indicates the size of the specifically amplified segment. The intense amplified segment below the 610 bp size standard (fourth largest size marker) is from an additional pair of primers that serve as an internal control for assessing the technical success of the PCR. PCR was performed with the following primers (from *figure 7*): lanes 1–6, primers 1 and 3; lanes 7–12, primers 2 and 3; lanes 13–18, primers 1 and 4; and lanes 19–24, primers 2 and 4. For each set of primers, the individuals are in order 1 thorugh 6. Lane S: standard, 250 ng φX174 *Hae*III restriction fragments. Lane N: no DNA template added as a control for contamination. From ref. 15 with permission.

A number of modifications of PASA have been developed to overcome these problems. A key issue has been the sensitivity of PASA under conditions where the frequency of the specific allele to be detected is less than 10^{-4} (e.g. presence of mutant oncogenes in tissues). However, in certain circumstances (e.g. carrier testing for genetic disorders), another important issue has been verification of the technical success of PASA in order to avoid false negatives.

5.1 Simultaneous detection of allele-specific products

One approach to verifying the technical success of PASA has been to perform both allele-specific amplifications in a single PCR. If at least one of the allele-specific PCR products is present, the technical success of the PASA is assumed. This approach does require the ability to distinguish between the two allele-specific PCR products.

Some of the modifications of PASA that allow simultaneous detection of both allele-specific PCR products include:

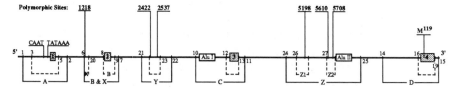

Figure 9. Schematic of the transthyretin (TTR) gene. Four exons (shaded boxes 1–4), the putative promoter, and two Alu repeat sequences are shown. Amplified regions are designated by solid lines, labelled A–D and X–Z. Sequenced regions are indicated by the broken lines within the amplified regions. The oligonucleotide primers used are shown with the downstream primers given above the line and the upstream primers below the line (numbers 1–23). The site of the M[119] mutation is shown, and the polymorphic sites are indicated by base pair numbers (1218, 2422, 2537, 5198, 5610, and 5708). From ref. 8 with permission.

Table 4. Haplotype analysis of the transthyretin (TTR) mutant M[119] allele by double PASA.[a]

	Allele at polymorphic sites[b]						
Patient	**1218**	**2422**	**2537**	**5198**	**5610**	**5708**	**TTR M[119]**
E109	T	G	G	A	C	G	Yes
	G	C	A	C	G	T	No
C171	T	G	G	A	C	G	–
	T	G	G	A	C	C	
C175	T	G	G	A	C	G	Yes
	G	C	A	C	G	T	No
C274	T	G	G	A	C	G	Yes
	G	C	A	C	G	T	No
S228	T	G	G	A	C	G	–
	T	G	G	A	C	C	

[a] For each patient, the haplotype in the top row is associated with the M[119] mutation as determined by double PASA with site 5610 and M[119] plus site 5708 and M[119].
[b] Polymorphic sites are as described in *Figure 9*.
Adapted from ref. 8 with permission.

(a) Using allele-specific oligonucleotides of different lengths in PCR amplification of multiple specific alleles (PAMSA).

(b) Differentially labelling the allele-specific oligonucleotides with fluorescent or radioactive tags.

(c) Performing bidirectional allele-specific amplification by tetra-primer PCR. Because both allele-specific oligonucleotides are present in the same PCR, the resulting competitive oligonucleotide priming can also increase the specificity of PASA (see Section 3.3).

5.1.1 PCR amplification of multiple specific alleles (PAMSA)

For PCR amplification of multiple specific alleles (PAMSA) (also termed allele discrimination by primer length, ADPL), one allele-specific primer is designed to be longer than the other allele-specific primer by adding 30 or more non-complementary bases (*Figure 10*) (11, 17). Because the additional bases will not hybridize to the DNA template, the T_m of the allele-specific primer will not be affected. The difference in size of the two resulting amplification products can be detected by 4% agarose gel electrophoresis, allowing both allele-specific primers to be used in one reaction tube. Thus, both heterozygotes and homozygotes can be detected in one PCR amplification reaction (*Figure 11*). Moreover, since an amplified product is always produced, no internal control is necessary.

The competition of the two allele-specific oligonucleotides in the PCR can also prevent high concentrations of template DNA from overwhelming specificity. The specificity of a set of primers used to detect a polymorphism in the factor IX gene could be overwhelmed by a five- to tenfold increase in template DNA concentration when used separately with PASA (11). By using PAMSA, however, specificity was retained for this set of primers even with a greater than 1000-fold increase in template DNA concentration.

We have successfully used PAMSA for specific amplification at three different sites (11). However, in one case the first allele-specific primer appeared to amplify more efficiently than the second allele-specific primer. It was, therefore, necessary to decrease the relative molar ratio of the two allele-specific primers 40-fold to obtain equal amplification. This may reflect the ability of an amplified product to serve as a primer for further cycles of PCR. Such 'megapriming' can artificially convert the shorter amplified segment to the larger size.

Another difficulty with PAMSA is that heteroduplexes may form between the allele-specific PCR products, making the bands difficult to distinguish. Under our conditions of electrophoresis, any heteroduplexes that formed between the two amplified products during the PAMSA of heterozygotes migrated at or near the homoduplexes. However, it is possible that up to four distinct bands could be detected.

5.1.2 Tetra-primer PCR

Tetra-primer PCR uses four oligonucleotides (two flanking and two internal) and two temperature programs (18). The mismatch is located in the middle of one of the internal primers. Otherwise, the two internal primers are complementary to each other. The flanking primers have a T_m at least 10 °C higher than the internal primers. In the first round of PCR, the annealing temperature is sufficiently high so that only the flanking primers generate a product. In the second round of PCR, the annealing temperature is reduced. The two internal oligonucleotides can hybridize to the opposite strands of the amplified product generated in the first round of PCR. The two internal oligonucleotides can then specifically amplify their respective alleles in this seminested PCR. The allele-specific elongation primed by these oligonucleotides will occur in opposite directions (i.e. bidirectionally). If the internal primers

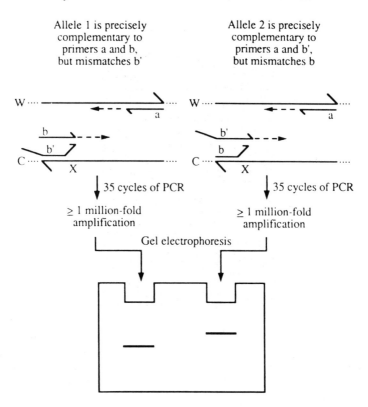

Figure 10. Schematic for polymerase chain reaction (PCR) amplification of multiple specific alleles (PAMSA). As in *Figure 1*, alleles 1 and 2 differ by a single base (represented by X where appropriate). Strands have been denatured by high temperature and the PCR oligonucleotide primers (a, b, and b') have been annealed. If the oligonucleotides are precisely complementary to the chromosomal DNA, elongation initiates efficiently. If there is a mismatch in the chromosomal DNA, elongation cannot be initiated efficiently. An allele-specific oligonucleotide, b, is designed to precisely match allele 1 and mismatch allele 2. A second allel-specific oligonucleotide, b', is designed to precisely match allele 2 and mismatch allele 1. The oligonucleotide b' also has 30 more non-complementary bases at the 5' end. In PAMSA, all three oligonucleotides (a, b, and b') are present during PCR. Since oligonucleotides a and b are complementary to allele 1, amplification of allele 1 will occur if present and a segment of appropriate size can be detected. No elongation of allele 1 will occur with oligonucleotide b'. Since oligonucleotides a and b' are complementary to allele 2, amplification of allele 2 will occur if present, and a segment 30 or more bases longer than the product from oligonucleotide a and b can be detected. No elongation of allele 2 will occur with oligonucleotide b. If both alleles 1 and allele 2 are present in the DNA sample, both alleles will be amplified, and the corresponding two segments can be separated and detected by gel electrophoresis. From ref. 10 with permission.

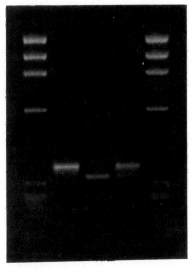

Figure 11. Specificty of PASMA. The polymorphism Alu 4a in X-linked factor IX gene (presence of C or T at nucleotide 28 653) was used to examine the specificity of PAMSA. Oligonucleotides were designed to specifically match either the C allele (primer 3) or T allele (primer 4). Primers 3 and 4 had identical complementary bases except for the 3′ ends containing G and A, respectively. Primer 4 also had 31 extra non-complementary bases to increase the length of the T allele amplified product. The Alu 4 repeat and unique flanking regions were initially amplified by PCR using primers 1 and 2. For PAMSA, the amplified products were then diluted 10 000-fold, and re-amplified with primers 1, 3, and 4. Lane S: standard, 250 ng φX174 *Hae*III restriction fragments. Lane 1: male with the T allele. Lane 2: male with the C allele. Lane 3: heterozygous female with one T allele and one C allele. From ref. 11 with permission.

are not in the centre of the original product, then their allele-specific products will be different lengths and can be distinguished by gel electrophoresis. In our hands, this method was not robust and offers little improvement in terms of specificity and sensitivity over PAMSA. However, it has been shown to be of value in the discrimination of alleles that differ by a nucleotide deletion or insertion (19).

5.2 Sensitivity of PASA

Methods such as PAMSA are useful for detecting mutations and polymorphisms in situations in which two alleles are commonly observed. However, PASA is better suited for simultaneously screening multiple samples for a rare allele. The sensitivity of PASA though is not always sufficient in cases in which the concentration of the rare allele is low relative to other templates. In these cases, various modifications of PASA have been exploited to increase the specificity and, hence, the sensitivity of PASA.

Some of the modifications developed to increase the sensitivity of PASA are:

(a) Adapting competitive oligonucleotide priming by employing two rounds of PCR in simplified allele-specific amplification (SASA).

(b) Using a competitive blocker in allele-specific competitive blocker PCR (ACB-PCR).

(c) Designing two or more mismatches in the allele-specific oligonucleotide in mismatch amplification mutation assays (MAMA) and mutagenically separated PCR (MS-PCR).

5.2.1 Simplified allele-specific amplification (SASA)

Simplified allele-specific amplification (SASA) avoids the accumulation of non-specific products by limiting the number of PCR cycles in two rounds of amplification (20). The first round generates the desired segment by competitive oligonucleotide priming (COP). The second round uses this segment as the target DNA for a low number of PCR cycles (< 15 cycles) so that unwanted non-specific products are not detected.

5.2.2 Allele-specific competitive blocker PCR (ACB-PCR)

Allele-specific competitive blocker PCR (ACB-PCR) is a variation of competitive oligonucleotide priming. One of the allele-specific oligonucleotides is labelled at the 3' end with a dideoxynucleotide to inhibit elongation (21). DNA products that span a mutation are generated by 15 cycles of asymmetric pre-PCR. This is followed with 25 cycles of semi-nested ACB-PCR using a mutation-specific oligonucleotide and a labelled oligonucleotide corresponding to the normal allele. The ACB-PCR results in a smaller fragment only if the mutant allele is present. The dideoxynucleotide labelled primer serves to block any false annealing of the mutation-specific primer to the normal alleles. The labelled primer might bind non-specifically to the mutant template, but amplification is prevented by the dideoxynucleotide. Thus, the competitive blocking may increase specificity and sensitivity.

5.2.3 Mismatch amplification mutation assay (MAMA)

Mismatch amplification mutation assay (MAMA) involves two mismatches instead of one mismatch in the allele-specific oligonucleotide to generate better specificity (22). The allele-specific oligonucleotide mismatches the mutant allele at one site and the normal allele at two sites. This double mismatch in the oligonucleotide prevented amplification of the normal allele even if it was present in great excess. These authors also observed that performing two-step PCR cycles gave better specificity than the typical three-step PCR cycles. The two-step PCR cycles included a denaturation step (94 °C for 1 min) and an annealing/elongation step (50°C for 1 min). The addition of glycerol (10%, v/v) further enhanced specificity.

5.2.4 Mutagenically separated PCR (MS-PCR)

Mutagenically separated PCR (MS-PCR) is a variation of PAMSA in which five mismatches are placed in the allele-specific oligonucleotides (23). The mismatches include the allele-specific mismatch at the 3' end plus four other mismatches (two near the 5' end and two near the 3' end of the primer). These four other mismatches differ between the two allele-specific primers and differ from the original target. Both of the allele-specific oligonucleotides are included in a single PCR reaction. Thus, several base differences are introduced into the allele-specific primer binding regions of the two corresponding PAMSA products. This reduces the cross-annealing of the PCR products with the wrong allele-specific primer in subsequent PCR cycles and therefore, increases specificity.

5.3 Other adaptations of PASA

PASA has been coupled to a variety of other techniques to improve:

(a) The efficiency of screening (e.g. multiplexing).

(b) The separation of the PCR products (e.g. capillary electrophoresis).

(c) The sensitivity of detection (e.g. electrochemiluminescence (ECL) and enzyme-linked immunosorbent assay (ELISA)-based detection).

However, PCR amplification of specific alleles is also easily amenable to adaptation for special requirements.

For instance, in the case of co-amplifying from single molecules, 'hemi-nesting' with ADPL (called PAMSA herein) has been successful (17). The initial round of amplification contained primers flanking the loci. The second hemi-nested amplification of the initial PCR contained one regular primer and two allele-specific primers of different lengths. By including three sets of primers, the haplotype of three independent genetic loci in single sperm samples could be determined.

PASA followed by single-strand conformation polymorphism (SSCP) instead of agarose gel electrophoresis can allow the detection of previously unknown single base sequence variations, small deletions, insertions, etc. PASA with SSCP can be used also to investigate complex genetic systems. For example, the human leucocyte antigen (HLA) loci have been analysed by PASA-SSCP to distinguish between the multiple series of HLA alleles (24).

Another variation of PASA used to study the HLA loci has been to couple the allele-specific oligonucleotides to streptavidin-coated magnetic beads. After PASA, the selectively amplified DNA can be then directly solid phase sequenced without cloning (25). These techniques serve to illustrate the power and adaptability of PASA when optimized.

6. Conclusions

PASA is a general method that can be optimized to detect all possible single base changes. The method may be also used to detect the presence of small deletions or insertions. PASA has the advantages of being rapid, reproducible, non-isotopic, adaptable, and amenable to automation.

References

1. Sommer, S. S., Cassady, J. D., Sobell, J. L., and Bottema, C. D. K. (1989). *Mayo Clinic Proc.*, **64**, 1361.
2. Sarkar, G., Cassady, J., Bottema, C. D. K., and Sommer, S. S. (1990). *Anal. Biochem.*, **186**, 64.
3. Wu, D. Y., Ugozzoli, L., Pal, B. K., and Wallace, R. B. (1989). *Proc. Natl. Acad. Sci.USA*, **86**, 2757.
4. Newton, C. R., Graham, A., Heptinstall, L. E., Powell, S. J., Summers, C., Kalsheker, N., *et al.* (1989). *Nucleic Acids Res.*, **17**, 2503.
5. Nichols, W. C., Liepnicks, J. J., McKusick, V. A., and Benson, M. D. (1989). *Genomics*, **5**, 535.
6. Okayama, H., Curiel, D. T., Brantly, M. L., Holmes, M. D., and Crystal, R. G. (1989). *J. Lab. Clin. Med.*, **114**, 105.
7. Sommer, S. S., Groszbach, A. R., and Bottema, C. D. K. (1992). *BioTechniques*, **12**, 82.
8. Ii, S., Sobell, J. L., and Sommer, S. S. (1992). *Am. J. Hum. Genet.*, **50**, 29.
9. Gustafson, S., Proper, J. A., Bowie, E. J. W., and Sommer, S. S. (1987). *Anal. Biochem.*, **165**, 294.
10. Bottema, C. D. K. and Sommer, S. S. (1993). *Mutat. Res.*, **288**, 93.
11. Dutton, C. and Sommer, S. S. (1991). *BioTechniques*, **11**, 700.
12. Gibbs, R. A., Nguyen, P. N., and Caskey, C. T. (1989). *Nucleic Acids Res.*, **17**, 2437.
13. Major, J. G. (1992). *BioTechniques*, **12**, 40.
14. Kwok, S., Kellog, D. E., McKinney, N., Spasic, D., Goda, L., Levenson, C., *et al.* (1991). *Nucleic Acids Res.*, **18**, 999.
15. Sarkar, G. and Sommer, S. S. (1991). *BioTechniques*, **10**, 436.
16. Lo, Y. M. D., Patel, P., Newton, C. R., Markham, A. F., Fleming, K. A., and Wainscoat, J. S. (1991). *Nucleic Acids Res.*, **19**, 3561.
17. Li, H., Cui, X., and Arnheim, N. (1990). *Proc. Natl. Acad. Sci. USA*, **87**, 4580.
18. Ye, S., Humphries, S., and Green, F. (1992). *Nucleic Acids Res.*, **20**, 1152.
19. Finckh, U., Rommelspacher, H., Schimdt, L. G., and Rolfs, A. (1996). *DNA Sequence*, **6**, 87.
20. Xu, L. and Hall, B. G. (1994). *BioTechniques*, **16**, 44.
21. Orou, A., Fechner, B., Utermann, G., and Menzel, H. J. (1995). *Hum. Mutat.*, **6**, 163.
22. Cha, R. S., Zarbl, H., Keohavong, P., and Thilly, W. G. (1992). *PCR Methods Appl.*, **2**, 14.
23. Rust, S., Funke, H., and Assmann, G. (1993). *Nucleic Acids Res.*, **21**, 3623.
24. Lo, Y. M. D., Patel, P., Mehal, W. Z., Fleming, K. A., Bell, J. I., and Wainscoat, J. S. (1992). *Nucleic Acids Res.*, **20**, 1005.

25. Kaneoka, K., Lee, D. R., Hsu, K. C., Sharp, G. C., and Hoffman, R. W. (1991). *BioTechniques*, **10**, 30.
26. Sommer, S. S., Sarkar, G., Koeberl, D. D., Bottema, C. D. K., Buerstedde, J. M., Schowalter, D. B., *et al.* (1990). In *PCR protocols: a guide to methods and applications* (ed. M. A. Innis, D. H. Glefand, J. J. Sninsky, and T. J. White), p. 197. Academic Press, San Diego.
27. Bottema, C. D. K., Koeberl, D. D., Ketterling, R. P., Bowie, E. J. W., Taylor, S. A. M., Lillicrap, D., *et al.* (1990). *Br. J. Haematol.*, **75**, 212.

11

The protein truncation test (PTT)

ROB B. VAN DER LUIJT, RICCARDO FODDE, and
JOHAN T. DEN DUNNEN

1. Introduction

The detection and characterization of point mutations in genes responsible for inherited disorders has become one of the most common practices in human molecular genetics. Several protocols are nowadays available: single strand conformation polymorphism (SSCP) (1), RNase protection (2), hydroxylamine and osmium tetroxide (HOT) chemical cleavage (3), and denaturing gradient gel electrophoresis (DGGE) (4, 5). These techniques are generally aimed at the identification of single base substitutions, insertions, or deletions in relatively small DNA fragments (50–500 bp). However, when dealing with large genes, it is desirable to analyse larger fragments of the coding region to speed up the analysis and pin-point the location of the alteration prior to the nucleotide sequence determination. To this aim, mRNA-based mutation detection protocols are often employed. Moreover, if the spectrum and the type (i.e. missense, nonsense, or frame-shift) of the prevalent mutations at a given disease-causing gene has already been established, one would prefer to employ a protocol 'targeted' for the detection of the most frequent type of mutations.

In this chapter we describe a mutation detection technique, the **p**rotein **t**runcation **t**est (PTT; (6–8)), aimed at the detection of any mutation truncating the protein product. The main strength of the PTT is its unique power to detect disease-causing mutations only and this in long, kilobase-sized fragments in one assay. PTT is based on *in vitro*-coupled transcription and translation of PCR-amplified coding sequences. We will describe the general aspects of the PTT protocol and give examples of DNA- and RNA-based applications to identify germline mutations in the *APC* (adenomatous polyposis coli) and *DMD* (Duchenne muscular dystrophy) genes respectively.

2. Mutation detection using PTT

The principle of the PTT is shown in *Figure 1*. In brief, templates for PTT are generated by PCR using either cDNA obtained through reverse transcription

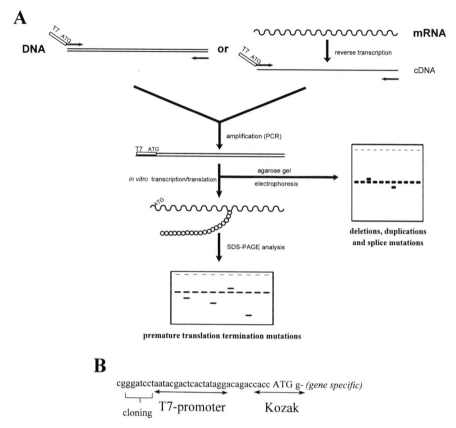

Figure 1. Principle of the Protein Truncation Test. (A) Schematic view of the PTT procedure starting from DNA or mRNA respectively. (B) Sequence of the tail added to the in-frame gene-specific primers.

(RT)-PCR of mRNA, or directly from genomic DNA. During PCR, a non-specific 5′ tail is added to the PCR product, which contains the bacteriophage T7 RNA polymerase promoter sequence, a translation initiation sequence, and an in-frame ATG triplet. The added tail allows *in vitro* coupled transcription and translation of the PCR products, performed in a single tube reaction. To facilitate detection of the *in vitro* synthesized polypeptides by autoradiography, radiolabelled amino acids are incorporated during translation. Translation products are resolved by SDS–PAGE and, after enhancement of the signal using a scintillating agent, detected by exposure to X-ray film. The appearance of a shortened translation product, compared to full-length polypeptides generated from normal control samples, indicates the presence of a truncating mutation in the gene segment analysed.

Depending on the genomic structure of the gene of interest, either DNA or RNA is used as starting material. If the target gene contains large, uninterrupted stretches of coding sequence, templates for the PTT can be generated by PCR on genomic DNA samples. Examples of genes with large exons are *APC* (Section 4.1), *BRCA1*, and *BRCA2* (*Table 1*). For the mutation analysis by PTT of genes which do not contain such large exons, a reverse transcription reaction is performed to synthesize a DNA copy of the messenger RNA (Section 4.2). For the DNA-based PTT protocol, genomic DNA can be obtained from peripheral blood lymphocytes by the salting out procedure (9).

Protocol 1. Isolation of DNA

Equipment and reagents

- Whole EDTA blood
- Centrifuge tubes, 50 ml
- Pasteur pipettes
- Erythrocyte lysis buffer: 155 mM NH$_4$Cl, 10 mM KHCO$_3$, 1 mM EDTA pH 7.4
- Saturated NaCl solution, 6 M

- NL buffer: 10 mM Tris–HCl pH 8.2, mM NaCl, 2 mM Na$_2$EDTA, 1% SDS, 200 μg/ml proteinase K
- TE buffer: 10 mM Tris–HCl pH 7.5, 0.1 mM EDTA
- Isopropanol (2-propanol)

Method

1. Transfer 10 ml of whole blood to a 50 ml centrifuge tube and mix the blood with a threefold volume of erythrocyte lysis buffer.

2. Incubate the mixture on ice for 30 min.

3. Centrifuge for 10 min at 2000 *g*.

4. Resuspend the white blood cell pellet in 10 ml NL buffer.

5. Incubate overnight at room temperature (alternatively, for 4–6 h at 37 °C).

6. Add 1/3 vol. of saturated NaCl and mix thoroughly by shaking.

7. Centrifuge for 15 *g* min at 2500 *g* to pellet the cellular proteins.

8. Collect the supernatant in a clean 50 ml tube and repeat step 7.

9. Using a blunt Pasteur pipette, transfer the supernatant to a clean 50 ml tube.

10. Add 1 vol. isopropanol. Mix gently by inverting the tube several times. The DNA should now become visible as long threads.

11. Using a Pasteur pipette with a bent tip, fish up the DNA from the solution.

12. Briefly wash the DNA in 70% ethanol.

13. Transfer the DNA to a 2 ml microcentrifuge tube containing 1 ml TE buffer (0.5 ml for small DNA pellets). Incubate the sample at 65 °C for 5 min to inactivate nucleases.

14. Allow the DNA to dissolve (for at least overnight).

15. Store the DNA samples at 4 °C.

Protocol 1. *Continued*

16. DNA prepared from 10 ml whole blood is dissolved in 1 ml TE buffer. In most cases, this results in a final DNA concentration of about 1 μg/μl, as determined by spectrophotometry

The quality of the RNA is a critical parameter in the RT-PCR based PTT procedure. RNA can be best isolated from a readily accessible source, i.e. from peripheral blood lymphocytes prepared from freshly drawn blood or from cultured cells. In *Protocol 2*, blood lymphocytes are prepared by density centrifugation using Histopaque-1077. There are several protocols for the isolation of RNA. We have obtained good results using the RNAzol™B method, which is a rapid and simple procedure. Always use sterile pipettes, tubes, and solutions. After RNA isolation, the quality and amount of the RNA are determined by analysis on an agarose gel.

Protocol 2. Isolation of RNA

Equipment and reagents

- Centrifuge tubes (5 ml and 10 ml)
- Sterile, RNase-free pipette tips and micro-centrifuge tubes (1.5–2 ml)
- Histopaque-1077 (Sigma)
- PBS: 144 mM NaCl, 10 mM KH_2PO_4, adjusted to pH 7.8 with K_2HPO_4
- DEPC treated solutions: add 0.05–0.1% (v/v) diethylpyrocarbonate (Sigma), leave overnight, and autoclave
- RNAzol™B (Cinna/Biotecx)
- Chloroform
- Isopropanol
- 70% ethanol

- TE buffer: 10 mM Tris–HCl pH 7.5, 0.1 mM EDTA
- Agarose (Seakem LE)
- 3 M NaAc: 408.1 g NaAc.$3H_2O$ dissolved in 1000 ml H_2O, adjusted to pH 5.6 with HAc
- 1 M NaOH
- Ethidium bromide: stock solution 10 mg/ml in H_2O
- RNA loading mix: 95% formamide, 20 mM EDTA, 0.05% bromophenol blue, 0.05% xylene cyanol
- TBE buffer: 90 mM Tris, 90 mM boric acid, 1 mM EDTA pH 8.3

A. *Preparation of blood lymphocytes*

1. Collect one tube (10 ml) EDTA blood.[a]

2. Fill two tubes (white caps) with 5 ml Histopaque-1077.

3. Slowly layer 5 ml of blood on top of the Histopaque.

4. Centrifuge at 950 *g* for 20 min at room temperature in a swing-out rotor (do not use brakes).

5. Four layers should be visible after centrifugation: the top layer contains serum and thrombocytes, the second layer (white) contains lymphocytes, the third layer (colourless) is the Histopaque, and the fourth layer contains the erythrocytes.

6. Using a Pasteur pipette, carefully remove and discard most of the first layer.

7. Collect the second layer and transfer it to a clean tube.

8. Add 10 ml cold and sterile PBS.

9. Centrifuge at 550 *g* for 10 min at room temperature.[b]

10. Remove the supernatant by carefully inverting the tube.

11. Resuspend the pellet in the remaining fluid by tapping the tube and put the cells on ice.

B. *Isolation of RNA*

Keep the RNA samples cold, wear gloves, and work with DEPC treated RNase-free solutions. Use RNase-free tubes and pipette tips.

1. Directly add 1.2 ml RNAzol™B to the cell suspension. A small white (sometimes red or reddish) pellet should be visible. Lyse the cells by passing the lysate several times through the pipette tip.

2. Transfer the lysate to a 1.5–2 ml microcentrifuge tube.

3. Add 0.1 ml chloroform per 1 ml of homogenate. Shake vigorously for 15 sec. Put the sample on ice for 5 min.

4. Centrifuge the suspension at 12 000 *g* for 15 min at 4 °C.

5. Two phases should be formed: an upper aqueous phase (colourless) and a lower phenol:chloroform phase (blue). Carefully transfer the aqueous phase (0.6–0.7 ml) to a clean tube. At this point, the solution can be stored at 4 °C until further processing.

6. Add an equal volume of isopropanol and place the sample on ice for 15 min.

7. Centrifuge at 12 000 *g* for 15 min at 4 °C. The RNA precipitate should be visible as a pellet at the bottom of the tube.

8. Remove the supernatant and wash the pellet with 200 μl 70% ethanol.

9. Centrifuge at 12 000 *g* for 10 min at 4 °C.

10. Carefully remove all supernatant and air dry the pellet (not too long).

11. Dissolve the pellet in 50 μl TE.

12. Precipitate the RNA by adding 5 μl 3 M NaAc pH 5.6 and 145 μl ethanol. Store the sample at −80 °C.

C. *Agarose gel electrophoresis*

1. Carefully clean the electrophoresis tank, the gel tray, and comb to remove any residual RNase activity. Incubate for 1 h with 1 M NaOH. Rinse extensively with water.

2. Pour a 1.5% agarose gel (Seakem LE) in sterile TBE and add ethidium bromide to 0.2 μg/ml.

3. Add to an aliquot of RNA sample (i.e. 5–10%) an equal volume of RNA loading mix.

Protocol 2. *Continued*

4. Incubate the sample for 5 min at 65 °C.

5. Load the sample on gel, together with a standard, providing a marker for size and amount.

6. Run the gel and inspect the RNA on a UV transilluminator. Comparison with the standard is used to determine the RNA concentration. The RNA quality is sufficient when the 18S and 23S ribosomal RNA bands are clearly visible and when no contaminating DNA band is observed.

[a] Alternatively, RNA can be isolated from cultured cells. Take a tissue culture flask (75 cm^2) with cells grown to 80–90% confluency. Discard the culture medium. Wash carefully with cold and sterile PBS. Add 5 ml PBS. Put the flask on ice and collect the cells by scraping with a sterile wiper. Transfer the cells to a clean tube. Add 5 ml PBS to the original flask and continue scraping, collecting as many cells as possible. Centrifuge the suspension at 550 *g* for 10 min. Carefully remove the supernatant by inverting the tube. Now proceed with part B of the protocol.

[b] For cells isolated from 5 ml blood or cultured cells from one 75 cm^2 flask, we use 1–1.3 ml RNAzol^{TM}B. The RNAzol^{TM}B contains chloroform and guanidinium chloride. It promotes the formation of complexes with RNA with guanidinium and water molecules and abolishes hydrophilic interactions of DNA and proteins, which are efficiently removed from the aqueous phase.

To provide a DNA target for the PCR reaction, cDNA is synthesized using a reverse transcription protocol.

Protocol 3. Reverse transcription

Equipment and reagents

- Random primers, 0.5 µg/µl (Promega)
- Superscript MMLV reverse transcriptase, 200 U/µl with RT buffer (BRL)
- RNasin ribonuclease inhibitor, 40 U/µl (Promega)
- RNaseH, 1.5 U/µl

- 10 mM dNTPs: 10 mM dATP, 10 mM dCTP, 10 mM dGTP, 10 mM dTTP (prepared from 100 mM stock, Pharmacia)
- 0.1 M DTT (BRL)
- TE buffer: 10 mM Tris–HCl, 0.1 mM EDTA pH 8.0

Method

1. Take 1–3 µg RNA (i.e. 50 µl ethanol stored RNA) in a sterile microcentrifuge tube and centrifuge at 12 000 *g* for 10 min at 4 °C.

2. Remove the ethanol using a sterile pipette and air dry the pellet. Do not dry the pellet for too long, otherwise it will not dissolve.

3. Add directly (or make a pre-mix and add): 2 µl random primer (0.5 µg/µl) and 30 µl TE. Mix, and incubate for 10 min at 65 °C. After the incubation, directly incubate the reaction on ice.[a]

4. Add directly (or make a pre-mix and add): 12 µl Superscript RT buffer,

6 μl 0.1 M DTT, 6 μl 10 mM dNTPs, 1 μl 40 U/μl RNasin, 2 μl 200 U/μl Superscript MMLV reverse transcriptase, and 1 μl sterile H_2O (the final volume is 60 μl). Mix, and place for 10 min at room temperature.

5. Incubate the reaction mixture for 60 min at 42 °C. Place the reaction on ice to stop the reaction.

6. Optional: add 4 U RNase H and incubate for 20 min at 37 °C.

7. When necessary, the procedure may be stopped at this point. If so, store the reaction mixture at −20 °C.

---

[a] Instead of random primers, gene-specific primers can be used to prime cDNA synthesis: scale down the amount of RNA to 200–500 ng and add 9.5 μl TE and 0.5 μl gene-specific reverse primer (20 pmol/μl). Anneal 10 min at 65 °C. Add (final volume 20 μl) 4 μl Superscript RT buffer, 2 μl 0.1 M DTT, 2.5 μl 10 mM dNTPs, 0.5 μl RNasin, and 1 μl 200 U/μl Superscript MMLV reverse transcriptase. Mix, place for 10 min at room temperature, and proceed with step 5.

For the RT-PCR based PTT procedure, a two-stage amplification reaction is often required to amplify the target sequences to a detectable level, using the first strand cDNA product (*Protocol 3*) as a template. In *Protocol 4*, steps 1–3 describe the first round of PCR, while steps 4–6 describe the second round PCR. If genomic DNA is used as starting material, steps 1–3 can be omitted.

Protocol 4. PCR using tailed primers

Equipment and reagents

- PCR apparatus
- AmpliTaq DNA polymerase, 5 U/μl (Perkin Elmer)
- Supertaq DNA polymerase (HT Biotechnology)
- Oligonucleotide primers: gene-specific (tailed) forward and reverse primers at 20 pmol/μl
- 10 mg/ml BSA (Boehringer)

- PCR buffer: 166 mM $(NH_4)_2$ SO_4, 67 mM Tris–HCl pH 8.8, 67 mM $MgCl_2$, 0.1 M β-mercaptoethanol
- Deep Vent DNA polymerase, 2 U/μl (Biolabs)
- 25 mM dNTPs: 25 mM dATP, 25 mM dCTP, 25 mM dGTP, 25 mM dTTP (prepared from 100 mM stock, Pharmacia)
- 100% DMSO

Method[a]

1. Prepare a pre-mix containing: 5 μl PCR buffer, 5 μl DMSO, 2.6 μl dNTPs (25 mM), 1 μl BSA (10 mg/ml), 0.4 AmpliTaq (5 U/μl), 0.04 μl Deep Vent (2 U/μl), and 22 μl H_2O.

2. Add 2 μl forward primer (20 pmol/μl), reverse primer (20 pmol/μl), 10 μl of the RT product, and two drops of mineral oil. Briefly spin down the reaction mixture.

3. Perform the PCR by subjecting the reaction mixture to an initial denaturation of 3 min at 93 °C, followed by 30 cycles of 1 min at 93 °C, 1 min at 58 °C, and 4 min at 72 °C. Finally, allow for an extension period of 7 min at 72 °C. Store the reaction product at 4 °C.

Protocol 4. *Continued*

4. Prepare a pre-mix containing: 5 μl Supertaq PCR buffer, 5 μl 2 mM dNTPs, 2 μl tailed forward primer (20 pmol/μl), 2 μl reverse primer (20 pmol/μl), 0.05 μl Supertaq, and 33 μl H₂O.

5. Add 3 μl of the first round PCR product (total volume is 50 μl). Overlay the reaction mixture with mineral oil.

6. Perform the PCR protocol as described in step 3.

[a] The total volume of the PCR can be scaled down to 25 μl (steps 1–3). PCR results can be improved by adding 0.1–0.2 U Deep Vent (step 3). For the DNA-based PTT procedure, perform PCR on 100–200 ng of genomic DNA (step 5).

Before the *in vitro* transcription/translation reaction is performed, the PCR products are analysed by agarose gel electrophoresis. This step is required to check the specificity of the amplification products and to determine the amount of template to be used in the *in vitro* coupled transcription/translation. Abnormal size of products generated by RT-PCR indicates the presence of genetic rearrangements (deletions, duplications, etc.) or mutations affecting RNA splicing. We use 0.5% agarose gels to analyse PCR products of 1–2 kb.

Protocol 5. Analysis of amplification products

Equipment and reagents
- Agarose gel electrophoresis unit
- UV transilluminator
- Agarose (electrophoresis grade)
- Ethidium bromide: 10 mg/ml in H₂O (stock solution)
- Electrophoresis buffer TBE (89 mM Tris, 89 mM boric acid, 2 mM EDTA, pH 8.3) TAE (40mM Tris-acetate, 1 mM EDTA, pH 7.6)
- DNA size marker

Method

1. Prepare a 0.5% agarose gel in electrophoresis buffer, containing 0.5 μg/ml ethidium bromide.

2. Place the gel in the electrophoresis chamber and fill the buffer compartments.

3. Load the gel with one-tenth of each PCR product and a DNA size standard.

4. Perform electrophoresis until the bromophenol blue dye reaches the bottom of the gel.

5. Remove the gel from the electrophoresis unit.

6. Inspect the gel under a UV transilluminator (wear safety glasses and gloves).

To produce translation products, the PCR fragments are transcribed using T7 RNA polymerase and the RNA is translated in a rabbit reticulocyte lysate.

The protocol is based on a one tube *in vitro* coupled transcription/ translation reaction: the TnT™ Coupled Reticulocyte Lysate System (Promega), using [^{3}H]leucine as a radiolabelled amino acid.

Protocol 6. Transcription and translation

Equipment and reagents

- TnT™ Coupled Reticulocyte Lysate System (Promega)
- RNasin ribonuclease inhibitor, 40 U/μl (Promega)
- L-(4,5)-[^{3}H] leucine, 5 μCi/μl (Amersham)
- Nuclease-free H$_2$O
- SDS sample buffer: 100 mM Tris–HCl pH 6.8, 4% SDS, 0.1% (w/v) bromophenol blue, 20% glycerol, and 8% (v/v) β-mercaptoethanol (added immediately before use)

Method

1. Remove the TnT reagents from the −70°C freezer. Place the TnT T7 RNA polymerase directly on ice. The RNA polymerase should not be kept outside the −70°C for too long. The reticulocyte lysate should be rapidly thawed by hand warming. After thawing, place all components on ice. Unused reticulocyte lysate should be refrozen as quickly as possible.

2. Pipette the following components in a 1.5 ml reaction tube: 12.5 μl TnT rabbit reticulocyte lysate, 1 μl TnT reaction buffer, 0.5 μl TnT T7 RNA polymerase, 0.5 μl Leu-free amino acid mix, 2 μl [^{3}H] leucine, 0.5 μl RNasin, and up to 8 μl of PCR product (50–500 ng). If necessary, adjust the final volume to 25 μl with nuclease-free H$_2$O.

3. As a control for the *in vitro* transcription/translation reaction, assemble a reaction containing 0.5 μl of the luciferase-encoding control plasmid supplied with the TnT kit.

4. Mix, and incubate the reaction for 60 min at 30°C.

5. Add 1 vol. of SDS sample buffer to the transcription/translation products and store the reactions at −70°C (or −20°C) until further use.

Purification of the PCR-products to be transcribed/translated is usually not necessary. Satisfactory results are also obtained by scaling down the reaction to a 12 μl volume with a ratio PCR product:TnT mix of up to 1:2. Reaction temperature can be varied between 25–37°C without greatly affecting translation efficiency, although 30°C seems to be optimal.

The *in vitro* synthesized products are resolved by gel electrophoresis using the discontinuous buffer system developed by Laemmli. The SDS–PAGE protocol presented here is based on the MiniProtean II gel system (Bio-Rad). This system requires only a short electrophoresis time (1–1.5 h) and uses small amounts of reagents to prepare and stain a gel. The gel described in *Protocol 7* can be used to resolve peptides from 15 kDa up to 70 kDa.

Protocol 7. SDS–PAGE analysis of translation products

Equipment and reagents

- Protein electrophoresis assembly (e.g. MiniProtean II gel system of Bio-Rad)
- Pre-stained protein M_r markers (Bio-Rad) or ^{14}C-labelled markers (Amersham)
- AA/BA mix: 30% acrylamide and 0.8% *N,N'-bis*-methylene acrylamide dissolved in a final volume of 500 ml, filtered, and stored in the dark at 4°C
- Separating gel mix, per 5 ml: 2 ml AA/BA mix, 1.65 ml distilled water, 1.25 ml 1.5 M Tris–HCl pH 8.8, 50 μl 10% SDS
- Stacking gel mix, per 4 ml: 0.5 ml AA/BA mix, 3 ml distilled water, 0.5 ml 0.5 M Tris–HCl pH 6.8, 40 μl 10% SDS
- APS: 10% ammonium persulfate in H_2O, freshly prepared
- Running buffer: 25 mM Tris base, 200 mM glycine, 0.1% SDS
- DMSO/PPO: 226 g 2,5-diphenyloxazole, dissolved in 1000 ml DMSO
- TEMED

Method

1. Add 50 μl 10% APS and 2 μl TEMED to 5 ml of 12% separating gel mix to start polymerization. Gently swirl the solution.

2. Pour the gel between the glass plates. Carefully overlay the gel with water or isobutanol (previously saturated with water) to obtain a sharp meniscus. Leave for at least 30 min to allow polymerization.

3. Rinse the top of the gel several times with water. Remove as much water as possible.

4. Take 2 ml of 3.75% stacking gel mix and add 20 μl 10% APS and 2 μl TEMED. Gently swirl the solution.

5. Pour the stacking gel on top of the separating gel and place a cleaned comb into the solution, carefully avoiding air bubbles trapped underneath the teeth of the comb. The stacking gel is ready in about 10 min.

6. Remove the comb and rinse the slots with running buffer. Place the gel in the electrophoresis tank and fill the chamber with running buffer.

7. Boil the samples for 5 min to denature the proteins. Centrifuge the samples for 30 sec in a microcentrifuge.

8. Load the gel with the samples, controls, and protein size markers.

9. Electrophoresis is performed at a constant current of 30 mA in the stacking gel and 40 mA in the separating gel. Stop the run when the bromophenol blue dye reaches the bottom of the gel. When translation products do not enter the gel (i.e. a strong signal appears in the slot), addition of 5% β-mercaptoethanol and omission of the boiling step may solve the problem. The luciferase control should produce a protein band at 62 kDa.

After electrophoresis, the gel is washed and a scintillating agent (PPO) is diffused into the gel to convert the energy emitted by the isotope to visible light, enabling detection of the translation products by exposure to X-ray film.

Protocol 8. Detection of translation products

Equipment and reagents
- Plastic trays for gel staining
- Vacuum slab gel dryer
- X-ray film, e.g. X-Omat AR (Kodak)
- Whatman 3MM paper
- Saran Wrap or similar
- DMSO
- PPO (226 g/litre) in DMSO
- Demineralized water

Method
1. Remove the gel from the electrophoresis assembly and place it in a plastic tray. Cover the gel with DMSO and wash for 10 min with gentle agitation.
2. Discard the washing solution. Repeat the DMSO washing step once.
3. Discard the DMSO and treat the gel twice with DMSO/PPO.
4. Remove the DMSO/PPO and wash the gel for 10–15 min with water.
5. Place the gel on 3MM paper and cover the gel with Saran Wrap. Dry the gel under vacuum at 60–70 °C for at least 1 h (depending on the thickness of the gel and the percentage of acrylamide used).
6. Expose with X-ray film.

DMSO, used to dehydrate the gel and enabling the subsequent infusion of PPO, is a hazardous organic chemical. Therefore, *Protocol 8*, steps 1–4 should be performed in a fume-hood. The DMSO/PPO solution can be reused two or three times. There are several alternative reagents for enhancement of the autoradiography, e.g. AMPLIFY (Amersham) and ENHANCE (New England Nuclear). Usually, the products obtained by *in vitro* transcription/translation of 50 ng of a 1.5 kb PCR products in a 12 µl reaction can be easily detected after overnight exposure.

3. Strategy

In general, as most human disease genes are too large to enable amplification as a single fragment, the coding region of the gene has to be split into overlapping segments. In our hands, segments of 1–2 kb in length proved to be most convenient to PTT analysis. Design of a segmented set requires careful attention. Choice of the primers ultimately determines the sensitivity obtained. Each segment is amplified using a tailed sense (or forward) primer and an antisense (or reverse) primer. To select the gene specific part of the primers, we use software packages like *PRIMER* (MIT, Cambridge, USA) and *OSP* (10).

Rob B. van der Luijt et al.

3.1 Forward primer

To allow *in vitro* transcription and translation, PCR is performed with a tailed forward gene-specific sense oligonucleotide, which introduces an RNA promoter and an in-frame translation initiation sequence at the 5' end of the amplified fragment (*Figure 1B*). The tailed primers we have used are based on the sequence described by Sarkar and Sommer (11), containing an 18 bp bacteriophage T7 promoter and an 8 bp eukaryotic translation initiation signal or Kozak sequence. Furthermore, to facilitate cloning of the amplified fragments, we have added a restriction endonuclease site to the 5' end.

3.2 Reverse primer

In general, we incorporate no translation termination codon in the reverse primer but generate the protein products by run-off translation from the amplified segments. We have once tested the effect of adding an in-frame termination codon but did not observe any improvement.

3.3 Segmented set

To construct a set of overlapping segments, it is essential that:

(a) The ATG initiation codon in the tailed sense primer is in-frame with the coding sequence. Otherwise, translation will start at the first internal sequence which contains an efficient translation initiation site and an unexpected, shorter translation product will be produced. To prevent that mutations affecting the natural translation initiation codon are not detected, the forward primer of the first segment should be located upstream of this site.

(b) Flanking segments contain an overlap ensuring detection of mutations close to their ends. In general, 150–200 bp overlaps (5–8 kDa after translation) for 1–2 kb segments should be sufficient. A forward primer should not be selected in a region where the sequence is devoid of a codon for the labelled amino acid to be incorporated. In such cases, the translation product is produced but not detected and the truncating mutation is missed. A reverse primer should not be selected near the end of a region where a large open reading frame is present in one or both of the alternative reading frames. If so, a frame-shifting mutation will not cause premature translation termination and it will be missed.

(c) When PTT is implemented for RNA-based mutation detection, primers employed to amplify flanking and/or overlapping segments are located in different exons. This is to avoid that due to an intragenic deletion of a single exon, or to a splicing defect, PCR fails to amplify both fragments of one allele. For the same reason, a primer which covers both the end of one and the beginning of the next exon should be avoided.

200

For genes which show differential splicing, PTT analysis can be improved by selecting the differentially spliced exons as a primer binding site. If so, PCR will amplify only full-length RNA molecules and not the differentially spliced forms. This simplifies the picture emerging after transcription/translation and SDS–PAGE since only one and not two protein fragments are produced.

The boundaries of the translated fragments represent the most critical regions for the PTT analysis: N terminal (early) mutations result in products which might be too small to allow detection (no or little label incorporated, electrophoretic migration outside the resolution range), C terminal (late) mutations might result in a size difference which is not resolved near the top of the gel. Mutations at translation initiation and termination sites represent a special case. Here, the internal control of the overlapping segment is not available. The emerin gene, responsible for Emery–Dreifuss muscular dystrophy (EMD), is an example of a gene where extreme N terminal mutations do occur (12): four different mutations have been identified which destroy the ATG initiation codon. An extreme C terminal mutation was found in a breast cancer case where a frame-shift mutation was identified only 11 amino acids before the natural translation termination site (13).

4. Examples

4.1 PTT using genomic samples: FAP

An example of DNA-based mutation detection by PTT is represented by the analysis of patients with familial adenomatous polyposis (FAP), an autosomal dominant colorectal cancer syndrome. FAP patients carry germline mutations in the adenomatous polyposis coli (*APC*) gene which invariably lead to the occurrence of premature stop codons and thus to the production of truncated proteins. The coding region of the *APC* gene measures 8.5 kb and is split into 16 exons. The most 3′ exon of *APC* is unusually large, encompassing 6.5 kb of contiguous coding sequence. The spectrum of constitutional *APC* mutations in FAP is extremely heterogeneous, with mutations scattered over almost the entire coding region. Using mutation detection techniques which permit the analysis of gene segments of 100–500 bp, such as DGGE and SSCP analysis, mutation analysis of *APC* is very laborious and time-consuming. Moreover, in comparison with DGGE and SSCP which require the coding region to be divided up in to 40 segments, PTT of *APC* requires only five segments: exons 1–14 are amplified as one segment from RNA, while the 6.5 kb exon 15 is amplified from genomic DNA in four overlapping segments (7, 8)

Figure 2 shows a PTT analysis of four FAP patients in which the translation terminating mutation is located at different positions in exon 15 of the *APC* gene. To allow mutation analysis of the segment covering codons 989–1700, a 2.1 kb template was prepared by PCR with a T7 modified *APC*-specific primer pair and genomic DNA as starting material (primer sequences and

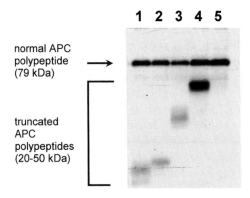

Figure 2. Example of the PTT. Analysis of a 2.1 kb segment of the *APC* gene amplified from genomic DNA (codons 989–1700). Lanes 1–4 contain translation products obtained from FAP patients with germline *APC* mutations at different positions within the *APC* gene. The sample in lane 5 is derived from a normal control. In each of the patients (but not in the control), a truncated APC protein (20–50 kDa) can be seen in addition to the full-length 79 kDa polypeptide.

PCR conditions as in ref. 14). In each case, a shortened polypeptide is seen in addition to the normal polypeptide of 79 kDa, which corresponds to the wild-type *APC* allele (note that all patients are heterozygous for the *APC* mutation). The length of the truncated proteins corresponds with the site of the stop codon in the amplified segment as determined by DNA sequencing. As it can be seen in lane 5, analysis of DNA from an unaffected control individual results in a single protein band, the full-length translation product.

An example of PTT analysis of the *APC* gene in members from one FAP family is shown in *Figure 3*. In the two affected persons (lanes 1 and 6), a shortened APC protein of 51 kDa is observed in addition to the normal 79 kDa polypeptide. Sequence analysis revealed a CGA→TGA mutation at codon 1450, substituting an arginine with a stop codon (15). Note that in each lane, several additional protein bands can be seen (indicated by asterisks). These signals are probably generated by secondary translation initiation at internal ATG codons in the amplified segment (see Sections 4.2 and 5.4).

4.2 PTT using RNA samples: DMD/BMD

Mutations in the human dystrophin gene cause Duchenne and Becker muscular dystrophy (DMD/BMD), a recessive X-linked disease. The gene is extremely complex; it is split into 79 exons which are spread over 2.4 Mb. In about two-thirds of the patients, large intragenic deletions and duplications are found with, as a general rule, out-of-frame mutations causing DMD and in-frame mutations causing the milder BMD phenotype. Detection of the deletion mutations is rather simple and straightforward using a multiplex PCR protocol which amplifies the 18 most frequently deleted exons (16). Due

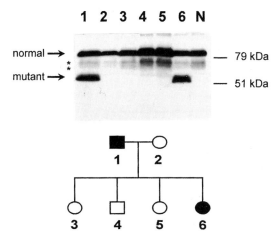

Figure 3. PTT analysis of the same segment as shown in *Figure 2*, performed on genomic DNA samples from members of one FAP family. Numbers above the lanes correspond to the individuals in the pedigree (affected persons are shown as filled symbols). In both patients, but not in the unaffected persons or the control sample (N), a truncated APC polypeptide of 51 kDa is seen, indicating the presence of a chain terminating mutation.

to the complexity of the gene, DNA-based methods have not yet been employed with great success to identify the (point) mutation in the remaining one-third of the cases. RNA-based methods, although more laborious, gave much better results (17). To scan the dystrophin gene for mutations with PTT, the coding region of the gene is amplified in 6–12 overlapping segments (17, 18), usually using blood-derived RNA as starting material (*Figure 4A*). Due to the extremely low expression of the dystrophin gene in this tissue, a two-step nested PCR protocol is required to amplify the gene product to detectable levels.

Since DMD is an X-linked disease, fragments amplified from the dystrophin gene of male patients are derived from one allele only. Consequently, when a translation terminating mutation is detected in a patient, no full-size normal translation protein is present (lane 2, *Figure 4B*). In contrast, analysis of carrier females does reveal a full-sized next to a truncated translation product (lane 2, *Figure 4C*). Since the truncated product in the DMD patient (lane 2, *Figure 4B*) is rather small (14 kDa), the mutation should be early in the amplified segment. Careful analysis of the autoradiograph shows that two background translation products migrate at 45 kDa in the patient as well as in the controls. Since these spurious products are not affected by the mutation they must have been derived from secondary translation initiation sites located downstream of the site of the mutation.

After amplification, the PCR products are first analysed on an agarose gel, both to estimate the amount of PCR product obtained and to check if the product has the expected length. Size alterations indicate the existence of

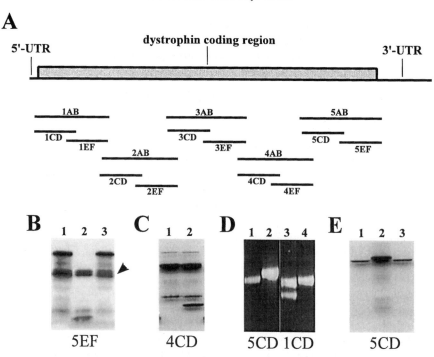

Figure 4. RNA-based PTT analysis in Duchenne muscular dystrophy. (A) The coding region of the dystrophin gene is split into ten overlapping segments. (B) Autoradiograph of translation products of segment 5EF of two controls (lanes 1 and 3) and a DMD patient (lane 2). (C) Autoradiograph of translation products of segment 4CD of a control (lane 1) and DMD carrier female (lane 2). (D) Agarose gel analysis (ethidium bromide stained) of PCR products derived from segments 5CD (lanes 1 and 2) and 1CD (lanes 3 and 4). Loaded are samples from two controls (lanes 1 and 4), a BMD patient (lane 2), and a DMD carrier female (lane 3). (E) Translation products derived from segment 5CD, i.e. the samples of lanes 1 and 2 of panel D together with that of a second control.

small deletions/duplications or the presence of mutations affecting splicing. *Figure 4D* gives an example of two such cases. Lanes 3 and 4 contain the PCR product of segment 1CD, amplified from a carrier female and a control respectively. Since the female is a carrier of a mutation abolishing splicing of one of the exons, two PCR products arise: one with and one without this exon. *Figure 4C*, lanes 1 and 2 contain PCR fragment 5CD of a normal control and BMD patient. It can be clearly seen that the product of the patient is slightly larger than that of the control. Translation of the product (*Figure 4E*) shows that the encoded protein is larger, indicating that the additional sequences are in-frame and do not introduce a translation termination signal. Sequence analysis of the cDNA showed the insertion of a 57 bp sequence between exons 64 and 65. Genomic sequence analysis of the patient revealed a G to C mutation at position 9569 + 5 of the splice donor site, causing the activation of a cryptic intronic splice site, 57 bp downstream of the original splice site (19). Since the reading frame remains intact, with a 19 amino acid insertion, the patient was expected to have a mild phenotype, i.e. BMD.

5. Troubleshooting and potential pitfalls

5.1 Contaminations

PTT mutation detection is for most genes performed using an RNA-based protocol on blood-derived RNA. Consequently, a nested PCR is required to amplify the target sequences, making the procedure extremely sensitive to contaminations of the sample and the ingredients of the RT-PCR reaction. Any previous PCR analysis is a major source of contamination. A 'pre-PCR' laboratory is thus essential to keep the RNA sample and primers physically separated from the RT-PCR and subsequent steps of the procedure. A control reaction (no RNA or no enzyme added) should always be part of the PTT to exclude contaminations.

5.2 No amplification

Due to the low amount of target, the nested RT-PCR procedure is very sensitive to an inefficient reverse transcription reaction and the RT reaction becomes the most critical step. Fresh constituents and fresh batches of enzyme are the best safeguards against emerging problems. Failure is often caused by the use of bad quality RNA. Freshness of the sample have a significant influence on success rate of the RT-PCR reaction. For blood samples we obtained best results with blood preserved in EDTA.

RNA samples derived from peripheral blood lymphocytes are usually suitable to amplify any gene of interest. In our hands, amplification of the *OTC* gene (20) was the only exception to this rule. In cases like this, buccal swabs and/or skin fibroblasts may provide alternative RNA source.

5.3 Side products after amplification

It is known that PTT-analysis of the 5' end of the BRCA1 and APC gene is difficult, mainly because more than one amplification product is obtained after RT-PCR. Such undesired products usually arise from alternative splice forms of the transcript and these products yield additional translation products. Therefore, it is wise to try alternative primer sets which give cleaner amplification products (see 3.3.c). PCR side products can also be derived from deletions or mutations affecting splicing. However, in general, such products can be easily recognized because they are unique for a specific sample.

5.4 No translation products

Occasionally, PCR products give poor yields in transcription/translation. This is often caused by a poor quality of the tailed primer. Inefficient coupling steps during primer synthesis result in low quantities of full-size tailed primers. Since primers are synthesized 3'→5', the T7 RNA polymerase promoter sequence is added last and most primers will then contain an incomplete T7 sequence. Such incomplete primers give normal yields in PCR, but *in vitro* transcription will fail. Re-amplification of the tailed products with a complete T7 promoter primer should solve the problem.

Translation products may require special treatments and/or running conditions to allow their visualization. For example, PTT of the exon 6–13 region of the cystic fibrosis (*CF*) gene, containing part of a transmembrane region, resulted in empty lanes (F. Hogervorst, personal communication). A strong signal near the top indicated that translation products were synthesized but did not enter the gel. Addition of 5% β-mercaptoethanol to the sample, and omission of the boiling step, solved the problem.

5.5 Background translation products/false positives

Upon translation, some sets produce a range of undesired background translation products, usually derived from secondary sites of translation initiation. Although these products could disguise truncated products, in general the translation pattern is fairly constant and changes are easy to detect. Furthermore, truncated bands from such sets go together with similar set of side-products, thereby simplifying their detection. To reduce background problems, a forward primer can be relocated or designed with more than one Kozak sequence. Alternatively, a wheat germ extract can be tried (21). Finally, using antibodies directed against a protein tag introduced with the tailed forward primer (22), it is possible to detect (or precipitate) correctly initiated translation products only.

Detection of a unique truncated fragment should always be verified by a second assay starting from an independent RT reaction. We have had a few exceptional cases where the truncated fragment could not be reproduced. The origin of these artefacts (false positives) is unclear but it probably resides in the infidelity of the reverse transcriptase and/or polymerases used to copy and amplify the RNA. Other false positives, i.e. a truncated translation product without a detectable mutation at the DNA level, have never been reported.

5.6 Missing mutations/false negatives

If existing translation terminating mutations are not resolved by PTT this usually originates from a failure to amplify the mutated allele. Failure of primer binding, either by deletion or mutation of the binding site, presents the most obvious example of such mutations. Amplification across (and beyond) translocation and inversion breakpoints will also yield no products. Problems may also derive from the intrinsic property of PCR to favour amplification of shorter products. Consequently, duplications will be detected only when they are small (below 1 kb exonic sequences) and when the primers used for amplification span the duplicated region. Similarly, insertions will be missed when the incorporated sequences go beyond a given size. The most convenient way to check for the potential presence of these genomic rearrangements is a Southern analysis of genomic DNA using cDNA probes. Mutations altering the splicing pattern are not detected when the mutated product increases beyond amplifiable sizes, e.g. when excision of an intron is abolished. Mutations in the promoter region result in lower levels of RNA from the affected allele: for BRCA1 a case was recently reported in which RNA-analysis yielded transcripts from only one allele (13).

Mutations causing premature translation termination have been reported to produce unstable RNA-molecules (23, 24), which should result in low (or no) amplification products of the mutated allele. However, although we have now studied many mutations in blood-derived RNA-samples, we have never observed such instability. This might be related to the fact that we amplify 'illegitimate transcripts', i.e. transcripts from a tissue where the protein is not required.

Expression of both alleles can be verified most efficiently using polymorphisms present in the coding region. For analysis of male patients with X-linked diseases, PTT results cannot be obscured by those of a healthy allele and even promoter mutations abolishing transcription are easily detected. However, analysis of females for X-linked diseases is further complicated by the interference of non-random X-inactivation and a potential positive selection of the healthy allele in somatic tissues. For autosomal diseases, expression of both alleles should be verified especially when no mutation could be identified.

Using RNA templates not derived from a disease-affected tissue, one should realize that the structure, i.e. the exon content, of the RNA can be different in different tissues. Consequently, mutations influencing tissue-specific expression, including the use of tissue-specific promoters and tissue-specific splicing, cannot be detected. Therefore, before mutation analysis is initiated, one should check if the RNA template to be analysed has the expected structure. In the case of the dystrophin gene it was reported that lymphocytic RNA contains an additional exon 'X', spliced between exons 1 and 2, which is not present in muscle RNA (25).

6. Conclusions

The protein truncation test or PTT (6), also known as the *in vitro* synthesized protein or IVSP (7) assay, is a method of direct mutation detection based upon the *in vitro* transcription and translation of PCR amplified sequences. PTT allows the selective detection of chain terminating mutations in large amplified segments and is uniquely useful for the molecular diagnosis of inherited conditions due to germline mutations which cause the truncation of the gene product. The method has two major advantages over other commonly employed protocols for direct mutation detection:

(a) Large coding segments (2–3 kb) can be screened in a single assay. In fact, the combination of PTT with the recently developed high-fidelity long-range PCR (26), will allow the analysis of even larger coding regions.

(b) PTT detects debilitating mutations only, i.e. mutations resulting in termination of the translation process. Silent polymorphisms and missense mutations will not be detected. Therefore, in contrast with virtually every other mutation detection protocol, sequence analysis is not strictly necessary to confirm the mutation.

From this point of view, mutation analysis of disease genes like *APC* or *BRCA-1*, whose mutation spectrum is characterized by a majority of truncating mutations, is greatly simplified. Furthermore, the size of the truncated protein provides a precise indication of the location of the stop codon within the corresponding segment, guiding sequence analysis towards the identification of the mutation at the DNA level.

Limitations of the PTT method for general use in human molecular genetics are:

(a) In its present form it cannot detect missense mutations. However, the combination of isoelectric focusing and PTT possibly represents the next step towards the extension of its detection range to a broader spectrum of mutations (6).

(b) Being, in most of its applications, an RNA-based technique, mutation analysis by PTT might be hampered by the limited availability of RNA samples and by their susceptibility to degradation by nucleases. In those diagnostic laboratories where DNA-based screening methods have so far been employed, the introduction of the PTT as a screening method may require resampling of individuals.

(c) The use of radioactively labelled amino acids. The development of non-radioactive protocols, commercially available from several suppliers, where proteins are detected by fluorescence or chemiluminescence has alleviated this problem.

In conclusion, the protein truncation test offers a valid alternative for the analysis of genes frequently inactivated by mutations predicted to truncate the protein product. It allows the analysis of large coding sequences or even entire genes with fewer PCR fragments than required for other commonly employed techniques. Truncating mutations have been shown to represent a very common cause of gene inactivation in a majority of inherited diseases (*Table 1*). Moreover, technical modifications of the PTT protocol can extend its applicability to missense mutations leading to charge alterations in the resulting polypeptide.

Table 1. Applications of PTT in human molecular genetics

Disease	truncating[a]	Gene	References
Duchenne muscular dystrophy	≥95%	*DMD*	(6,17)
Familial adenomatous polyposis	≥95%	*APC*	(7,8)
Hereditary desmoid disease	100%	*APC*	(27)
Hereditary breast- and ovarian cancer	≥90%	*BRCA1*	(28)
Hereditary breast cancer	≥90%	*BRCA2*	*(29)*
Hereditary non-polyposis colorectal cancer	~80%	*hMSH2*	(30)
	~70%	*hMLH1*	(31)
Hunter syndrome	~50%	*IDS*	(32)
Neurofibromatosis type 1	≥50%	*NF1*	(33)
Neurofibromatosis type 2	≥65%	*NF2*	(34)
Polycystic kidney disease	95%	*PKD1*	(35)
Rubinstein–Taybi syndrome	10%	*RTS*	(36)

[a] The percentage of truncating mutations reported which should be detectable using PTT.

References

1. Orita, M., Iwahana, H., Kanazawa, H., Hayashi, K., and Sekiya, T. (1989). *Proc. Natl. Acad. Sci. USA*, **86**, 2766.
2. Myers, R. M., Larin, Z., and Maniatis, T. (1985). *Science*, **230**, 1242.
3. Cotton, R. G. H., Rodrigues, N. R., and Campbell, R. D. (1988). *Proc. Natl. Acad. Sci. USA*, **85**, 4397.
4. Myers, R. M., Fischer, S. G., Lerman, L. S., and Maniatis, T. (1985). *Nucleic Acids Res.*, **13**, 3131.
5. Fodde, R. and Losekoot, M. (1994). *Hum. Mutat.*, **3**, 83.
6. Roest, P. A. M., Roberts, R. G., Sugino, S., Van Ommen, G. J. B., and Den Dunnen, J. T. (1993). *Hum. Mol. 2, Genet.*, 1719.
7. Powell, S. M., Petersen, G. M., Krush, A. J., Booker, S., Jen, J., Giardello, F. M., *et al.* (1993). *N. Engl. J. Med.*, **329**, 1982.
8. Van Der Luijt, R. B., Meera Kahn, P., Vasen, H., Van Leeuwen, C., Tops, C., Roest, P. A. M., *et al.* (1994). *Genomics*, **20**, 1.
9. Miller, S. A., Dykes, D. D., and Polesky, H. F. (1988). *Nucleic Acids Res.*, **16**, 1215.
10. Hillier, L. and Green, P. (1991). *PCR Methods Appl.*, **1**, 124.
11. Sarkar, G. and Sommer, S. S. (1989). *Science*, **244**, 331.
12. Yates, J., Aksmanovic, V., McMahon, R., Bione, S., and Toniolo, D. (1996). *Eur. J. Hum. Genet.*, **4**, 62.
13. Shattuck-Eidens, D., McClure, M., Simard, J., Labrie, F., Narod, S., Couch, F., *et al.* (1995). *J. Am. Med. Assoc.*, **273**, 535.
14. Van Der Luijt, R. B., Hogervorst, F. B. L., Den Dunnen, J. T., Meera Kahn, P., and Van Ommen, G. J. B. (1996). In *Laboratory protocols for mutation detection* (ed. U. Landegren), p. 140. Oxford University Press, Oxford.
15. Van Der Luijt, R. B., Meera Khan, P., Vasen, H. F. A., Tops, C. M. J., Van Leeuwen-Cornelisse, I. S. J., Wijnen, J. T., *et al.* (1997). *Hum. Mutat.*, **9**, 7.

16. Beggs, A. H. (1994). In *Current protocols in Human Genetics*, (ed. N. C. Dracopoli, J. L. Haines, B. R. Korf, D. T. Moir, C. C. Morton, C. E. Seidman, *et al.*). p. 9.3.1. New York: John Wiley & Sons, Inc.
17. Gardner, R. J., Bobrow, M., and Roberts, R. G. (1995). *Am. J. Hum. Genet.*, **57**, 311.
18. Roest, P. A. M., Van Der Tuijn, A. C., Ginjaar, H. B., Hoeben, R. C., Hogervorst, F. B. L., Bakker, E., *et al.* (1996). *Neuromusc. Disord.*, **6**, 195.
19. Roest, P. A. M., Bout, M., Van Der Tuijn, A. C., Ginjaar, H. B., Bakker, E., Hogervorst, F. B. L., *et al.* (1996). *J. Med. Genet.*, **33**, 935.
20. Grompe, M., Muzny, D. M., and Caskey, C. T. (1989). *Proc. Natl. Acad. Sci. USA*, **86**, 5888.
21. Hope, I. A. and Struhl, K. (1985). *Cell*, **43**, 177.
22. Ahn, A. H. and Kunkel, L. M. (1995). *J. Cell Biol.*, **128**, 363.
23. Baserga, S. J. and Benz, E. J. Jr. (1988). *Proc. Natl. Acad. Sci. USA*, **85**, 2056.
24. Dietz, H. C., Valle, D., Francomano, C. A., Kendzior, R. J. Jr., Pyeritz, R. E., and Cutting, G. R. (1993). *Science*, **259**, 680.
25. Roberts, R. G., Bentley, D. R., and Bobrow, M. (1993). *Hum. Mutat.*, **2**, 293.
26. Cheng, S., Fockler, C., Barnes, W. M., and Higuchi, R. (1994). *Proc. Natl. Acad. Sci. USA*, **91**, 5695.
27. Eccles, D. M., Van Der Luijt, R. B., Breukel, C., Bullman, H., Bunyan, D., Fisher, A., *et al.* (1996). *Am. J. Hum. Genet.*, **59**, 1193.
28. Hogervorst, F. B. L., Cornelis, R. S., Bout, M., Van Vliet, M., Oosterwijk, J. C., Olmer, R., *et al.* (1995). *Nature Genet.*, **10**, 208.
29. Lancaster, J. M., Wooster, R., Mangion, J., Phelan, C. M., Cochran, C., Gumbs, C., *et al.* (1996). *Nat Genet.*, **13**, 238.
30. Liu, B., Parsons, R. E., Hamilton, S. R., Petersen, G. M., Lynch, H. T., Watson, P., *et al.* (1994). *Cancer Res.*, **54**, 4590.
31. Papdopoulos, N., Nicolaides, N. C., Wei, Y. F., Ruben, S. M., Carter, K. C., Rosen, C. A., *et al.* (1994). *Science*, **243**, 1625.
32. Hogervorst, F. B. L., Van Der Tuijn, A. C., Poorthuis, B., Kleyer, W., Bakker, E., Van Ommen, G. J. B., *et al.* (1994). *Am. J. Hum. Genet.*, **55**, A223.
33. Heim, R. A., Kam-Morgan, L. N. W., Binnie, C. G., Corns, D. D., Cayouette, M. C., Farber, R. A., *et al.* (1995) *Hum. Mol. Genet.*, **4**, 975.
34. Pulaski, K., Pettingell, W., Ward, C. J., and Gusella, J. F. (1995). *Am. J. Hum. Genet.*, **55**, A237.
35. Roelfsema, R. H., Spruit, L., Ward, C. J., Van Ommen, G. J. B., Peters, D. J. M., Harris, P. C., *et al.* (1994). *Am. J. Hum. Genet.*, **55**, A240.
36. Petrij, F., Giles, R. H., Dauwerse, J. G., Saris, J. J., Hennekam, R. C. M., Masuno, M., *et al.* (1995). *Nature*, **376**, 348.

Functional assay of the p53 tumour suppressor gene

THIERRY FREBOURG, JEAN-MICHEL FLAMAN,
ANNE ESTREICHER, and RICHARD IGGO

1. Introduction

Screening methods generally used to detect mutations are commonly based on structural analysis of genes. These methods are often labour-intensive and fail to distinguish inactivating mutations from polymorphisms or functionally silent mutations. The alternative is to use biological assays which test the function of the gene product. This chapter describes a functional assay which detects mutations in human p53 cDNA expressed in the yeast *Saccharomyces cerevisiae*. Somatic p53 mutations are the most common genetic defect known to occur in human tumours (for a review see ref. 1) and more than 2000 different mutations have been described. Germline p53 mutations predispose to the development of multiple tumours and are responsible for the majority of cases of the familial Li–Fraumeni cancer syndrome.

The most straightforward approach to detect somatic p53 mutations is immunohistochemistry. The underlying assumption is that when p53 is mutant it is overexpressed, but this approach is limited by both biological and technical factors. Biological sources of confusion include false negatives due to nonsense mutations, false positives due to activation of wild-type p53 by DNA damaging agents, and unclassifiable tumours containing only small numbers of p53 positive cells. Technical sources of confusion arise from large variations in the way samples are prepared, particularly differences in fixation technique and use of microwave antigen retrieval, and use of antibodies with markedly different sensitivity. Detection of p53 mutations by DNA sequencing is expensive and labour-intensive. Screening methods such as SSCP, DGGE, and chemical mismatch cleavage have successfully been used to detect germline and somatic p53 mutations, but since they also detect polymorphisms it is essential to sequence abnormally migrating bands. The p53 functional assay described in this chapter can be used rapidly to detect either germline mutations in blood samples, and somatic mutations in tumours and cell lines.

2. Detection of human p53 mutations in yeast

The critical biochemical function of p53 which underlies its tumour suppressor activity is the ability to activate transcription (1). Mutant proteins fail to activate transcription in mammalian cells (2). The transcriptional activity of human p53 is conserved in yeast, and mutants which are inactive in human cells are also inactive in yeast (3). Detection of p53 mutations by testing the transcriptional competence of human p53 expressed in *Saccharomyces cerevisiae* is attractive given the advantages of low cost, ease of manipulation, and efficiency of cloning by gap repair in this organism (4, 5).

3. Overview of the assay

Budding yeast are co-transformed with PCR amplified p53 cDNA and a linearized expression vector, and the p53 cDNA is cloned *in vivo* by homologous recombination (*Figure 1*) (3–5). Since the vector expresses p53 from the ADH1 promoter, recombinants constitutively express low levels of human p53 protein. The recipient yeast strain (yIG397) is defective in adenine synthesis because of a mutation in the endogenous ADE2 gene, but it contains a second copy of the ADE2 open reading frame controlled by a p53-responsive promoter (5). Since ADE2 mutant strains grown on low adenine plates turn red, yIG397 colonies containing mutant p53 (or no p53) are red, whereas colonies containing wild-type p53 are white (5). Since adenine is limiting for growth, red colonies are slightly smaller than white ones.

To test a clinical sample, the following steps are required: mRNA extraction, cDNA synthesis, PCR amplification, yeast transformation. Subsequently, yeast are grown for two or three days at 35°C and p53 status is determined by counting red and white colonies (*Figure 1*).

3.1 mRNA extraction

Budding yeast can not splice mammalian introns, so it is not possible to test genomic DNA. The use of mRNA requires special conditions for sample collection and storage, and, unfortunately, most routinely processed samples, including paraffin-embedded formalin-fixed tissue, can not be used. The most efficient way to prevent RNA degradation is to place samples directly in RNA lysis buffer at the point of collection. Thus, the understanding and cooperation of clinicians is the most important single factor affecting the outcome of the assay. mRNA is preferred to total RNA, which can sometimes give high backgrounds. Many protocols are available for mRNA purification, and these can be used without modification for abundant sources of high quality RNA such as cell lines. Since the amount of RNA in clinical samples is often limiting, we present a protocol employing oligo(dT) coupled to magnetic beads which allows analysis of small samples such as endoscopic biopsies. To screen

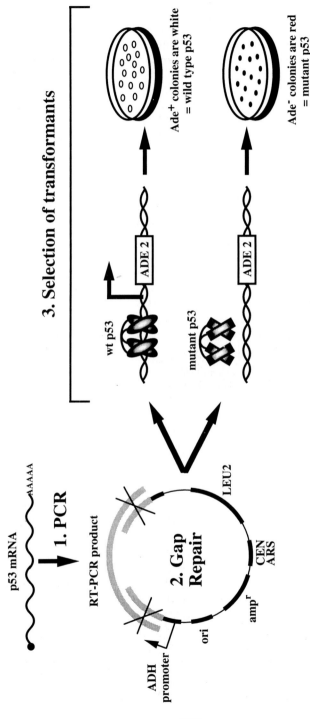

Figure 1. Outline of the assay. 1. p53 mRNA is reverse transcribed and amplified by PCR. 2. The PCR product is co-transformed into yeast with a linearized expression vector carrying the 5' and 3' ends of the p53 open reading frame. Gap repair of the plasmid with the PCR product results in constitutive expression of human p53 protein. 3. Yeast which have repaired the plasmid are selected on media lacking leucine. The media contains sufficient adenine for growth of Ade⁻ cells, but they form red colonies. Hence colonies containing wild type p53 are white (*ADE2*⁺) and colonies containing mutant p53 are red (*ade2*⁻).

for germline p53 mutations we recommend isolation by Ficoll gradient of lymphocytes from 10 ml of freshly drawn blood. Failure to separate lymphocytes promptly leads to a progressive increase in background due to the accumulation of an alternatively spliced form of the p53 mRNA (6) (J. M. F. and T. F. unpublished data).

Protocol 1. Poly(A) mRNA purification

Equipment and reagents[a]

- MPC-E magnet (Dynal, 120.04)
- Oligo(dT) coupled to magnetic beads (Dynabeads) (Dynal, 610.05)
- Lysis buffer: 500 mM LiCl, 1% LiDS, 100 mM Tris–HCl pH 8.0, 10 mM EDTA, 5 mM DTT
- Elution buffer: 2 mM EDTA pH 8.0
- LiDS buffer: 150 mM LiCl, 0.1% LiDS, 10 mM Tris–HCl pH 8.0, 1 mM EDTA
- Wash buffer: 150 mM LiCl, 10 mM Tris–HCl pH 8.0, 1 mM EDTA
- RNase inhibitor (Boehringer, 799.017)
- Homogenizer (Polylabo, 06180)
- Rotating wheel

Method

1. Place samples as soon as possible in 500 μl lysis buffer and store at ≤ –20°C.

2. Transfer 75 μl of Dynabeads per sample to a 1.5 ml microcentrifuge tube and let it stand for 30 sec in an MPC-E magnet. Remove supernatant and wash the beads once with lysis buffer.

3. Homogenize the sample in the lysis buffer with a hand-held homogenizer or by passing it repeatedly through a 22G needle and freeze–thawing with dry ice.

4. Spin in a microcentrifuge for 1 min at full speed to pellet insoluble material.

5. Add the samples to the Dynabeads and place on a rotating wheel for 20 min.

6. Place the tubes in MPC-E magnet for 1–2 min and remove supernatant.[b]

7. Wash the beads twice with 500 μl LiDS buffer and once with wash buffer.

8. Add 20 μl elution buffer and incubate for 2 min at 65°C. Place tube in magnet for 1 min. Transfer the supernatant into a new tube containing 1 μl RNase inhibitor. Store frozen at ≤ –20°C.

[a] A Dynabeads mRNA direct kit (Dynal, 610.11), can also be used. This is expensive but reduces the risk of PCR contamination. It is also possible to use a QuickPrep Micro mRNA Purification kit (Pharmacia, 27–9255–01). In this case the RNA is eluted in 200 μl. Do not precipitate the RNA to concentrate it (this can give high backgrounds).
[b] Keep the supernatant, since it contains genomic DNA which can be used to confirm the presence of mutations.

3.2 cDNA synthesis

Three choices exist for priming cDNA synthesis: oligo(dT), random hexamers, and p53-specific oligos. The p53 mRNA has a long 3′ untranslated
region, which rules out use of oligo(dT). Hexamers can be used, but if the
mRNA is slightly degraded the use of hexamers can give rise to artefactual
intragenic deletions in the amplified p53 cDNA, presumably by a PCR
splicing mechanism. We therefore recommend use of a specific primer, and
we use a dodecamer primer because this is more efficient than using longer
primers.

Protocol 2.   p53 mRNA reverse transcription

Reagents
- Superscript II reverse transcriptase (Life Technologies, 18064–022)
- RT-1 primer (Genset): CGG GAG GTA GAC at 50 μg/ml

Method
1. Add 10 μl of RNase-free water to 5 μl of mRNA.[a] Heat at 65°C for 10 min, then chill on ice.

2. Prepare the reaction mix:

    - 4 μl   5 × first strand buffer
    - 2 μl   0.1 M DTT
    - 1 μl   dNTPs (10 mM each)
    - 2 μl   RT-1[b]
    - 0.5 μl   Superscript II reverse transcriptase (100 U)

3. Add 10.5 μl mRNA to the reaction mix.

4. Incubate for 1 h at 46°C.

5. Store at –80°C.

[a] Use more mRNA and less water when very small biopsies are tested.
[b] Random hexamers (pd(N)$_6$, Pharmacia, 27–2166) at 0.2 μg/ml can also be used.

3.3 PCR amplification of p53 cDNA

The extent to which PCR introduces mutations is widely underestimated.
Amplification of a 1 kb target with *Taq* polymerase through 35 doublings
introduces mutations in about 70% of the product DNA molecules. For most
uses this is not a problem, but the p53 gene is uniquely sensitive to point
mutation, and since mutations at about 500 sites will inactivate the protein,
the use of *Taq* polymerase gives about 35% background red colonies in
the assay (7). It is therefore essential to use a high fidelity PCR polymerase
for the functional assay. Among commercially available thermostable DNA

polymerases, Pfu currently has the highest fidelity, and we recommend using it for the p53 assay (7). This gives a maximum background due to PCR infidelity of 3.5% after 35 cycles. The proofreading 3′ to 5′ exonuclease activity of Pfu polymerase can lead to degradation of primers, resulting in non-specific priming and poor yield. To prevent this we use primers with a single phosphorothioate linkage at the 3′ end.

Protocol 3. PCR amplification of p53 cDNA

Reagents

- Pfu polymerase 2500 U/ml (Stratagene, 600.159)
- Primers at 50 µg/ml: -s- means 3′ terminal phosphorothioate linkage
- P3 primer (Genset): ATT TGA TGC TGT CCC CGG ACG ATA TTG AA-s-C
- P4 primer (Genset): ACC CTT TTT GGA CTT CAG GTG GCT GGA GT-s-G

- P16 primer (Genset): GCG ATG GTC TGG CCC CTC CTC AGC ATC TT-s-A
- P17 primer (Genset): GCC GCC CAT GCA GGA ACT GTT ACA CA-s-T
- 10 mM dNTPs (Pharmacia, 27–2035–01)

Method

1. Prepare a master reaction mix, which contains for each sample:
   - 2 µl   5′ primer[a]
   - 2 µl   3′ primer[a]
   - 2 µl   10 × Pfu buffer
   - 2 µl   DMSO
   - 0.1 µl   dNTPs
   - 9.4 µl   sterile distilled water

2. Place 17.5 µl of reaction mix in a PCR tube and add a drop of mineral oil.

3. Add 2 µl cDNA and 0.5 µl Pfu polymerase. Keep tubes on ice.

4. Pre-heat block to 95°C before placing tubes in it ('hot block start').

5. Do 35 PCR cycles using the following conditions (PE DNA Thermal Cycler): 94°C for 30 sec, 65°C for 60 sec, 78°C for 80 sec.[b]

[a]Choice of primers depends on which gap repair vector will be used (see below): for pRDI-22 use P3 and P4; for pFW34 use P4 and P16; for pFW35 use P3 and P17.
[b]The background due to self-ligation of the gap repair vector falls as the ratio of PCR product to vector rises. 35 cycles are therefore better than 30 cycles when the yield is low (five cycles only add 0.5% to the background from PCR infidelity). If the PCR yield is low after 35 cycles, suspect a problem with RNA quality and do not proceed with the assay.

---

4. Yeast transformation

yIG397 should routinely be cultured on complete medium supplemented with 200 µg/ml adenine to avoid selection of spontaneous suppressors of the endogenous mutant *ade2* locus. To test p53 status, yeast are plated on synthetic minimal medium minus uracil, minus leucine, plus 2.5–5 µg/ml adenine. Never reuse strains that have been plated on low adenine (they will give confusing results because the low adenine conditions of the assay are enough to select spontaneous ADE2 suppressors). Never maintain stocks of the strain by streaking it sequentially from plates kept at 4°C. Instead, always go back to a frozen stock that is known to have the correct genotype: mat **a** *ade2-1 leu2-3, 112 trp1-1 his3-11, 15 can1-100 ura3-1 URA3 3xRGC::pCYC1::ADE2*.

Cloning by gap repair in yeast is straightforward: unpurified PCR product is mixed with a linear yeast vector carrying the 5′ and 3′ end of p53 cDNA and transformed into yeast by standard lithium acetate transformation. The vector carries the LEU2 gene to allow selection of recombinants on plates lacking leucine; unrecombined linear DNA is rapidly lost. In theory the yeast could become leu$^+$ by self-ligation of the vector or by gene conversion of the endogenous *leu2-3, 112* locus, but in practice this does not occur to a significant extent. Three gap repair vectors are available (8): pRDI-22 is used routinely (the gap lies between codons 67 and 346); pFW35 tests codons 67 to 210; and pFW34 tests codons 211 to 346. The pFW 'split assay' vectors are useful in doubtful cases, such as tumours giving under 20% red colonies, because clonal mutations should be positive with only one of the vectors; if both vectors give similar numbers of red colonies suspect poor quality RNA causing intragenic deletions (8).

Protocol 4. Yeast transformation

Equipment and reagents

- Yeast strain yIG397
- Yeast expression vectors: pRDI-22, pFW34, pFW35
- Yeast media (see below): YPDA^{++} liquid, SDA^{5}HTαα plates
- 10 × LiOAc (1 M pH 7.5): 10.2 g LiOAc in 90 ml distilled water, adjust pH with acetic acid, then adjust volume to 100 ml—filter sterilize
- 10 × TE: 0.1 M Tris–HCl pH 7.5, 10 mM EDTA (sterilize by autoclaving)
- 50% PEG: 50 g PEG 4000 in 100 ml distilled water (sterilize by autoclaving)
- 'LiOAc/TE': mix 1 ml 10 × LiOAc, 1 ml 10 × TE, and 8 ml sterile distilled water
- 'LiOAc/TE/PEG': mix 1 ml 10 × LiOAc, 1 ml 10 × TE, and 8 ml 50% PEG

- Carrier DNA: single-stranded sonicated salmon sperm DNA at 10 mg/ml
- 3 mm glass beads (Polylabo, 52415)
- YPDA^{++} (1 litre):10 g yeast extract (Difco, 0127–17–9), 20 g peptone (Difco, 0118–17–0), 20 g dextrose, 40 ml adenine (0.5%) (for plates, add 25 g bacto agar to 1 litre of medium)
- SD (1 litre): 6.7 g yeast nitrogen base w/o amino acids (Difco, 0919–15–3), 20 g dextrose
- SDA^{5}HTαα: to 1 litre of SD (cooled to 50°C), add 1 ml adenine (0.5%), 5 ml Trp (1%), 5 ml His (1%), 5 ml αα stock solution
- αα stock solution (100 ml): 0.4 g Arg, 0.4 g Met, 0.6 g Ile, 1 g Phe, 2 g Glu, 2 g Asp, 3 g Val, 4 g Thr, 8 g Ser—filter sterilize

Method

1. Inoculate 10 ml YPDA^{++} with one colony of yIG397 and grow overnight at 30°C.

Protocol 4. *Continued*

2. Dilute about 5 ml overnight culture into 50 ml pre-warmed YPDA^{++} to give an OD_{600} of 0.2 and grow at 30°C to an OD_{600} of 0.8–1.0 (usually takes 4–5 h).[a]

3. Transfer the culture to a 50 ml plastic tube and centrifuge.

4. Wash at room temperature with 50 ml sterile water, then 10 ml sterile water, then resuspend in 5 ml LiOAc/TE.

5. Centrifuge and resuspend pellet in 250 µl LiOAc/TE.[b]

6. Heat carrier DNA to 100°C for 5 min, then place on ice.

7. For each transformation, prepare one tube with:
   - 5 µl carrier DNA
   - 1 µl gap repair vector[c]
   - 5 µl unpurified RT-PCR product

8. Add 50 µl yeast in LiOAc/TE from step 5 and flick the tube to mix. Add 300 µl LiOAc/TE/PEG, vortex briefly, and incubate at 30°C for 30 min in a shaking incubator.

9. Heat shock at 42°C for 15 min (avoid agitation of the samples at this stage).

10. Spin briefly and resuspend the pellet in 200 µl sterile distilled water. Spread 20 µl and 180 µl of the cell suspension on SDA^{5}HTαα plates using glass beads.[d]

11. Incubate for two days at 35°C, then place at 4°C. Count red and white colonies on a plate with 100–300 colonies in total.[e]

[a] Do not let the cells grow beyond an OD of 1.0. If they overgrow the efficiency will drop.

[b] Once the cells are in LiOAc they can be left at room temperature for several hours without ill effects. 250 µl yeast in LiOAc/TE (from a 50 ml culture) is enough for about ten transformations.

[c] Each set of transformations should include linearized gap repair vector alone: it should give less than five colonies in total. To prepare the gap repair vector digest pRDI-22 with *Hind*III and *Stu*I, digest pFW34 and 35 with *Sma*I and *Bgl*II; treat with calf intestinal alkaline phosphatase, extract with phenol:chloroform, precipitate with ethanol, and resuspend in sterile distilled water at 5 ng/µl. Check the background with every new batch of gap repair vector (uncut plasmid will give a high background). To transform yeast with circular plasmid DNA, use 5 µl carrier DNA and 100 ng plasmid.

[d] Wash 3 mm diameter glass beads in ethanol, then distilled water, then bake at 200°C for 30 min. To obtain evenly spread colonies put transformation on plate, add about ten beads per plate, close the lid, and shake vigorously for 30 sec. About ten plates can be shaken at a time. This gives more evenly spread colonies than traditional spreaders, and is less trouble.

[e] If there are too many colonies the red ones may be overgrown by the white ones. The colour gets stronger after a few days at 4°C.

5. Interpretation of results

Clinical material containing wild-type p53 commonly gives a background of 5–10% red colonies using the conditions we describe, and higher values are indicative of the presence of p53 mutations (5). The background is due to polymerase errors during the transcription, reverse transcription and PCR re-actions (4, 7), and to the presence of an alternatively spliced p53 mRNA encoding a truncated protein with a transcriptional defect (6). Germline mutations should give around 50% red colonies, but the exact ratio depends on the relative amounts of wild-type and mutant mRNA. Tumours containing somatic p53 mutations may give a mixture of red and white colonies because of the presence of contaminating normal tissue (5). Some p53 mutants give pink colonies: for clinical purposes these should be scored as mutant. Pink colonies may contain temperature-sensitive mutants—to test this plate at 35 °C and 25 °C (5). If two different mutant alleles are present they can recom-bine to recreate the wild-type sequence (5). If you know that a sample is clonal, the presence of low numbers of white colonies (e.g. 15%) is thus a clue that the cells may be heterozygous for two different mutations.

Careless sample preparation can give serious artefacts. If blood samples are left for 24 hours before separating the lymphocytes, the background due to alternative splicing of intron 9 can reach 50% (J. M. F. and T. F. unpublished data). If tissue is not processed quickly, RNA degradation may occur, leading to high backgrounds from intragenic deletions. Since these deletions are not clonal, it is possible to identify them with the pFW 'split assay' vectors.

Some mutations may be missed. These include promoter mutations, large deletions, and nonsense mutations. Nonsense mutations in the 5′ part of the message lead to RNA degradation ('nonsense-mediated mRNA decay') which can mask the presence of mutations. This is rarely a problem with p53 because most mutations are point missense mutations, but, as with all tests, a negative result should be interpreted in the context of the clinical data or the biological problem being addressed.

In our experience, red colonies virtually always contain plasmids with p53 mutations. Thus it is not necessary to perform additional analyses in straight-forward cases. At times, however, it may be necessary to identify the exact mutation present. This can easily be done by rescuing plasmids from red colonies. To avoid misleading results due to PCR mutations it is sensible to sequence one plasmid from at least three different red yeast colonies. Many protocols are available for rescuing plasmids from yeast.

Protocol 5. Rescuing plasmids from yeast colonies

Reagents

- Wizard Minipreps DNA Purification System (Promega, A 7100)[a]
- Zymolyase (Seikagaku Corporation, Tokyo, Japan, 120491) 4 mg/ml in 1 M sorbitol (stable at 4 °C)

Protocol 5. *Continued*

Method

1. Grow a 10 ml yeast culture in YPDA^{++} to an OD$_{600}$ of about 1.

2. Pellet and resuspend in 500 μl 1.2 M sorbitol, 20 mM EDTA pH 8.0.

3. Add 50 μl zymolyase and incubate for 60 min at 37 °C.

4. Pellet the cells (spheroplasts) and purify DNA according to the Wizard Minipreps DNA Purification standard protocol.[b]

5. Precipitate the DNA using glycogen as carrier, resuspend in 5 μl water, and electroporate DH^{5}α.

6. To check the vector cut with *Bst*YI. Expect bands of 4.5, 1.6, 1.1, 1.0, and 0.8 kb for the prototypic cDNA, and bands of 5.5, 1.6, 1.1, and 0.8 kb for the alternately spliced form (intron 9 retention).

[a]This is based on a standard bacterial alkaline lysis DNA miniprep. The reason for using the Wizard affinity column is that it removes inhibitors of bacterial transformation.
[b]The solution should become viscous after addition of the NaOH/SDS solution. If not, the cell wall was not sufficiently digested. Pre-incubation of the yeast in reducing agents will improve lysis: at stage 2 resuspend pellet in 1 ml 0.1 M Tris–HCl, 0.1 M EDTA pH 8.0, add 50 μl of β-mer-captoethanol, and incubate for 10 min at room temperature.

6. Conclusion

The yeast assay described here can be used to test peripheral blood lympho-cytes for germline p53 mutations (5). It can also be used to detect p53 mutations in tumours, and our experience with head and neck tumours and brain tumours indicates that the yeast approach is substantially more sensitive than conventional techniques (8) (M. Tada and R. I. unpublished data). Many factors probably contribute to this high sensitivity:

(a) The assay tests a larger region of the p53 cDNA than is normally tested (amino acids 53–364 versus amino acids 126–306 in most conventional studies).

(b) Analysis of cDNA means splicing defects are readily detected.

(c) Mutations can be detected in the presence of large amounts of normal tissue.

(d) Cloning artefacts are not a problem because hundreds of clones are examined from each sample.

(e) Biologically important mutations are correctly identified on the basis of their biological activity.

(f) The simple red/white read-out means that mutations are not easily over-looked.

Acknowledgements

A. E. and R. I. are supported by the Swiss National Science Foundation and the Swiss Cancer League. J. M. F. and T. F. are supported by La Ligue Nationale Contre le Cancer, l'Association pour la Recherche sur le Cancer, Le Groupement des Enterprises Francaices dans la Lutte Contre le Cancer. Requests for strains and plasmids should be addressed to R. I.

References

1. Ko, L. J. and Prives, C. (1996). *Genes Dev.*, **10**, 1054.
2. Frebourg, T., Barbier, N., Kassel, J., Ng, Y. S., Romero, P., and Friend, S. H. (1992). *Cancer Res.*, **52**, 6976.
3. Scharer, E. and Iggo, R. (1992). *Nucleic Acids Res.*, **20**, 1539.
4. Ishioka, C., Frebourg, T., Yan, Y. X., Vidal, M., Friend, S. H., and Iggo, R. (1993). *Nature Genet.*, **5**, 124.
5. Flaman, J. M., Frebourg, T., Moreau, V., Charbonnier, F., Martin, C., Chappuis, P., *et al.* (1995). *Proc. Natl. Acad. Sci. USA*, **92**, 3963.
6. Flaman, J. M., Varidel, F., Estreicher, A., Vannier, A., Limacher, J-M., Gilbert, D., *et al.* (1996). *Oncogene*, **12**, 813.
7. Flaman, J. M., Frebourg, T., Moreau, V., Charbonnier, F., Martin, C., Ishioka, C., *et al.* (1994). *Nucleic Acids Res.*, **22**, 3259.
8. Waridel, F., Estreicher, A., Bron, L., Flaman, J. M., Fontolliet, C., Monnier, P., *et al.* (1997). *Oncogene*, **14**, 163.

13

Advances in direct DNA sequencing for mutation scanning

JOAKIM LUNDEBERG, CECILIA WILLIAMS,
ASFHIN AHMADIAN, and MATHIAS UHLÉN

1. Introduction

The rate of discovery for disease-causing genes and underlying mutations is accelerating, and a number of genes, such as the tumour suppressor gene p53, cumulatively affect large numbers of individuals in the general population. Screening for single nucleotide changes and less frequent frame-shift losses/ gains of one to a few nucleotides is most readily accomplished by direct DNA sequencing. The transformation of DNA sequencing from research to clinical laboratories will probably lead to routine procedures for prognostic and therapeutic genes in specialized laboratories within ten years (1, 2). The primary issues to be addressed for applying direct DNA sequencing as a clinical tool include automation, optimal methods to obtain high quality data, efficient multiplex amplification stategies, quality control, and procedures to deal with very small samples.

The introduction of the polymerase chain reaction (PCR), the replacement of isotopic labels by fluorescent dyes, and on-line monitoring of the DNA sequencing results has significantly contributed to that, and automated DNA sequencing has gained attention as an attractive possibility for routine genetic analysis. The use of PCR for amplification of genetic material allows for direct DNA sequencing, and so avoids tedious cloning steps and the obligatory control sequencing necessary to exclude 'mutations' introduced by the *Taq* DNA polymerase (3). Thus the DNA sequence of a sample can be determined rapidly and will represent the sequence of the sample prior to amplification, as the *Taq* DNA polymerase errors will not significantly contribute to the resulting signal. Various methods can be employed to obtain labelled dideoxy DNA fragments according to the Sanger methodology (4) such as using labelled primers or dideoxy chain terminators or nucleotides. The detection principle of most commercial DNA sequencing instruments relies on the excitation of these fluorescent labelled sequencing bands by a laser beam during electrophoresis, enabling quantitation of polymorphic and heterozygous positions using sophisticated software tools (5–8).

2. Solid phase sequencing

Solid phase technology (i.e. magnetic beads) has contributed to standardizing DNA sequencing since reaction buffers and additional reagents can be rapidly changed without semi-quantitative steps such as centrifugation or precipitation. The manual protocols for magnetic bead assisted DNA sequencing described below have been used in the process of developing automated DNA sequencing systems. The template preparation and the sequencing reactions have been implemented on a Beckman Biomek 1000 workstation (9, 10) and on an ABI Catalyst workstation (11).

The concept of direct solid phase DNA sequencing involves superparamagnetic beads coated with streptavidin used to immobilize and purify biotinylated PCR products, and to generate single-strand templates (12). The approach takes advantage of the strong ($K_d = 10^{-15}$ M) and temperature stable (up to 80 °C) features of the biotin and streptavidin interaction, which is also tolerant to alkali treatment. After amplification with a biotinylated primer, the biotinylated PCR products are efficiently immobilized onto streptavidin-coated magnetic beads within minutes and the captured DNA can subsequently be denatured into two separate strands by the addition of sodium hydroxide. This results in the elution of the non-biotinylated strand into the supernatant, while the biotinylated strand remains immobilized to the bead surface. The single-stranded template in the supernatant can be recovered by the use of a magnet and subsequent removal of the eluate into a separate tube for neutralization. The advantage of the method is that all reaction components are removed, including the complementary strand, enabling optimal sequencing conditions with no reannealing problems. The DNA polymerase from phage T7 has generally been used for DNA sequencing, due to its good processivity and low discrimination in incorporation nucleotide analogues when substituting Mn^{2+} for Mg^{2+}. The DNA polymerase from *Thermus aquaticus* is also a useful enzyme, especially in cycle sequencing. However, *Taq* DNA polymerase incorporates nucleotide analogues with varying efficiency resulting in variation of the signal intensity and is therefore not suitable for heterogeneous clinical samples. However, recent manipulations of the existing polymerases have generated modified enzymes such as AmpliTaq DNA polymerase CS (Perkin Elmer) and Thermo Sequenase DNA polymerase (Amersham) with improved properties making them more suitable for analysis of mixed sequences.

The use of solid phase direct sequencing in polymorphic analysis has been documented for human genetic diseases (13–15), as well as bacterial 16S RNA operons (16), and virus populations (5–7, 17, 18). A high degree of sequence heterogeneity is frequently found in tissue biposy samples due to the presence of surrounding normal cells and in retroviral samples due to the error prone nature of the RNA-directed RNA polymerase of the virus. Thus, to study these samples in a population-based approach, such as direct sequencing, is well suited and actually simplifies the handling of a sample as compared with the frequently used method of sequencing a number of individual PCR clones.

3. Solid phase DNA sequencing for p53 mutation scanning

A detailed procedure for sequence-based analysis of the human p53 gene is presented below and the protocols can easily be adapted to other targets and instruments. To facilitate the investigation of the scattered mutations in the human p53 gene we have modified the standard DNA sequencing protocol both regarding chemistry (dNTPα) as well as sample handling (multiplex PCR) (19). First, a multiplex PCR approach is employed, where exons 4 to 9 and the HLA-DQB1 locus are co-amplified simultaneously. The polymorphic sequence of the HLA-DQB1 locus is included to allow for sample identification and as a contamination control. Secondly, solid phase sequencing is performed by using fluorescent labelled primers and Sanger methodology, where deoxynucleotides (dNTP) are replaced by deoxy α-thionucleotides (dNTPα), and the subsequent sequence analysis is performed by an automated laser fluorescent instrument. The introduction of α-thionucleotides resolves regions with secondary structures, and the improved sequence quality enables accurate determination of p53 alterations and HLA polymorphisms.

3.1 Sample preparation

Fresh tumour tissues are snap-frozen for subsequent manual microdissection under the microscope. Tumour nests containing less than 50 cells can be isolated from 16 μm cryostat sections stained with methylene blue and/or p53-specific antibodies (13). The cells are lysed and then used for outer multiplex PCR.

Protocol 1. Sample preparation

Reagents
- Sample solution: 10 mM Tris–HCl pH 8.3 (20°C), 50 mM KCl, 0.1% Tween 20
- Proteinase K (20 mg/ml)

Method
1. Transfer the microdissected tumour cells to tubes containing 50 μl sample solution.
2. Cover with light mineral oil.

Protocol 1. *Continued*

3. Add 2 µl proteinase K and incubate at 56°C for 1 h.

4. Inactivate proteinase K by incubation at 95°C for 5 min. Put on ice.

5. Transfer 5 µl of the mixture to PCR tubes containing outer PCR reaction mixtures (see *Protocol 5*).

3.2 Nested multiplex *in vitro* amplification

DNA in the the microdissected samples are amplified in a multiplex PCR approach developed for amplification of seven separate gene regions. This allows for a common single tube amplification of all gene fragments, thus reducing the total required number of cells necessary for the PCR. Exons 4 to 9, covering the mutational hot spot regions in the human p53 gene (20), and the HLA-DQB1 locus are co-amplified using the 14 outer PCR primers presented in *Table 1*. All primers are situated in intronic sequences. The labelled oligonucleotides are synthesized with a biotin amidite or a fluorescent label FITC (fluorescein isothiocyanate) amidite at their 5′ ends, and finally purified by reverse-phase chromatography. After outer amplification, the multiplex mixture is divided into separate tubes for region-specific amplification (see *Figure 1*). The PCR parameters have been optimized in regard to primer composition, primer concentration, inner cycle number, Mg^{2+} concentration, sample volume, annealing temperatures, and DNA polymerase employed, in order to achieve consistent amounts of inner fragments, see *Figure 2*. The nested primer approach enables multiplexing at low stringent conditions in the outer amplification (50°C) without affecting the quality of the inner fragments. This will enable uniform amplification also of only a few starting DNA molecules. To accomplish reproducible amplification of the GC-rich exon 5 a mixture of two DNA polymerases are used, AmpliTaq® DNA polymerase and Stoffel DNA polymerase. Without addition of the Stoffel fragment in both the outer and inner amplifications of exon 5, the amount of final PCR product is unpredictable. The Stoffel fragment has an increased thermal stability which allows for amplification, with higher denaturation temperature (98°C), of regions with strong secondary structures.

Protocol 2. Multiplex amplification

Equipment and reagents

- 10 × PCR buffer: 100 mM Tris–HCl pH 8.3 (20°C), 20 mM $MgCl_2$, 500 mM KCl, 1% Tween 20
- 10 × PCR buffer (for exon 5, inner PCR): 100 mM Tris–HCl pH 8.3 (20°C), 25 mM $MgCl_2$, 10 mM KCl, 1% Tween 20
- Perkin Elmer 9600 thermocycler (Perkin Elmer Cetus, California, USA)

- AmpliTaq® DNA polymerase and AmpliTaq® DNA polymerase Stoffel fragment (Perkin Elmer Cetus, Norwalk, CT, USA)
- 10 × dNTP nucleotide solution: 2 mM of each dNTP
- PCR primer solutions[a]: 5 µM of each primer

Method

1. Prepare an outer PCR master mix in a microcentrifuge tube: 5 μl 10 × PCR buffer, 1 μl of each primer solution (14 primers), 5 μl dNTP solution, 1 U of AmpliTaq® DNA polymerase, 2 U of AmpliTaq® DNA polymerase Stoffel fragment, and sterile water to 45 μl. Multiply by number of samples. Aliquot into PCR tubes.

2. Cover with light mineral oil (50 μl).

3. Add cell lysate (*Protocol 1*).

4. Outer PCR program: a single denaturation step, 94°C 5 min, followed by 30 cycles comprised of 94°C 30 sec, 50°C 30 sec, 72°C 1 min. End by final extension at 72°C 10 min.

5. Prepare region-specific inner polymerase chain reaction master mix in a microcentrifuge tube.

    (a) For exons 4, 6–9, and HLA: 5 μl 10 × PCR buffer, 1 μl of each primer solution (two primers), 5 μl 10 × dNTP solution, 1 U Ampli-Taq® DNA polymerase, and sterile water to 48 μl for exons 4, 7, and 8, and to 49.5 μl for exons 6 and 9.

    (b) For exon 5: 5 μl 10 × PCR buffer, 1 μl of each primer solution (two primers), 5 μl dNTP solution, 2 U of AmpliTaq® DNA polymerase, Stoffel fragment, and sterile water to 48 μl. Aliquot into PCR tubes.

6. Cover with light mineral oil.

7. Add 2 μl template (outer PCR), for exon 6 and 9 add 0.5 μl template.

8. Inner PCR cycle:

    • For exon 4: 94°C 30 sec, 60°C 30 sec, 72°C 1 min for 30 cycles

    • For exon 5: 98°C 15 sec, 63°C 30 sec, 72°C 45 sec for 30 cycles

    • For exon 6–9: 94°C 30 sec, 63°C 30 sec, 72°C 1 min for 30 cycles

    • For HLA: 94°C 30 sec, 50°C 30 sec, 72°C 1 min for 30 cycles

    Each of the inner amplifications is preceded by a denaturation step at 94°C for 5 min except exon 5 (98°C for 3 min), and followed by a final extension step at 72°C for 10 min.

9. Analyse the PCR product on an agarose gel stained with ethidium bromide. Amplification of p53 and HLA results in the fragment sizes indicated in *Figure 2*.

---

[a] Biotinylated oligonucleotides should be purified from unbound biotin, preferably by reverse-phase FPLC or HPLC, as free biotin will occupy binding sites on the beads and reduce the binding capacity of biotinylated PCR products.

Table 1. Oligonucleotides for multiplex amplification and sequencing

Region	Primer	Nucleotide position	Sequence
Exon 4	HEAN 71[a]	11946–11965	5′-CTGGGACCTGGAGGGCTGGG-3′
	BECE 4[b]	11978–11998	5′-**biotin**-CTGAGGACCTGGTCCTCTGAC-3′
	BECE 5[b]	12332–12351	5′-ATACGGCCAGGCATTGAAGT-3′
	BECE 3[a]	12372–12391	5′-AGAGGAATCCCAAAGTTCCA-3′
	HEAN 75[d]	11985–12001	5′-**FITC**-CCTGGTCCTCTGACTGC-3′
	HEAN 76[c]	12314–12331	5′-**FITC**-CTCATGGAAGCCAGCCCC-3′
Exon 5	RIT 596[a]	12952–12967	5′-GGAGGTGCTTACACAT-3′
	RIT 597[b]	13007–13025	5′-**biotin**-TTCACTTGTGCCCTGACTT-3′
	HEAN 55[b]	13255–13274	5′-ACCAGCCCTGTCGTCTCTCC-3′
	HEAN 56[a]	13276–13295	5′-GAGGCCTGGGGACCCTGGGC-3′
	HEAN 65[d]	13021–13040	5′-**FITC**-GACTTTCAACTCTGTCTCCT-3′
	HEAN 66[c]	13251–13267	5′-**FITC**-CTGTCGTCTCTCCAGCC-3′
Exon 6	HEAN 77[a]	13251–13270	5′-GGCTGGAGAGACGACAGGGC-3′
	HEAN 78[b]	13274–13293	5′-**biotin**-TTGCCCAGGGTCCCCAGGCC-3′
	HEAN 45[b]	13453–13472	5′-CTTAACCCCTCCTCCCAGAG-3′
	HEAN 50[a]	13475–13494	5′-CGGAGGGCCACTGACAACCA-3′
	HEAN 79[b]	13286–13302	5′-**FITC**-CCCAGGCCTCTGATTCC-3′
	HEAN 46[c]	13453–13472	5′-**FITC**-CTTAACCCCTCCTCCCAGAG-3′
Exon 7	HEAN 57[a]	13939–13958	5′-TGCTTGCCACAGGTCTCCCC-3′
	HEAN 58[b]	13963–13982	5′-**biotin**-CGCACTGGCCTCATCTTGGG-3′
	HEAN 59[b]	14128–14147	5′-CAGCAGGCCAGTGTGCAGGG-3′
	HEAN 60[a]	14149–14168	5′-CGGCAAGCAGAGGCTGGGGC-3′
	HEAN 67[d]	13968–13984	5′-**FITC**-TGGCCTCATCTTGGGCC-3′
	HEAN 68[c]	14124–14140	5′-**FITC**-CCAGTGTGCAGGGTGGC-3′
Exon 8	HEAN 61[a]	14395–14414	5′-ACAGGTAGGACCTGATTTCC-3′
	HEAN 62[b]	14420–14439	5′-**biotin**-GCCTCTTGCTTCTCTTTTCC-3′
	HEAN 63[b]	14609–14628	5′-CCCTTGGTCTCCTCCACCGC-3′
	HEAN 64[a]	14630–14649	5′-TGAATCTGAGGCATAACTGC-3′
	HEAN 69[d]	14425–14444	5′-**FITC**-TTGCTTCTCTTTTCCTATCC-3′
	HEAN 70[c]	14600–14616	5′-**FITC**-TCCACCGCTTCTTGTCC-3′
Exon 9	HEAN 80[a]	14612–14632	5′-GTGGAGGAGACCAAGGGTGC-3′
	HEAN 81[b]	14638–14660	5′-**biotin**-GCCTCAGATTCACTTTTATCACC-3′
	RIT 599[b]	14777–14796	5′-CTGGAAACTTTCCACTTGAT-3′
	RIT 598[a]	14891–14906	5′-GTTAGCTACAACCAGG-3′
	HEAN 82[d]	14652–14670	5′-**FITC**-TTTATCACCTTTCCTTGCC-3′
	RIT 474[c]	14777–14796	5′-**FITC**-CTGGAAACTTTCCACTTGAT-3′
HLA-DQB1	DB130[a, b]		5′-AGGGATCCCCGCAGAGGATTTCGTGTACC-3′
	DB131[a]		5′-TCCTGCAGGGCGACGACGCTCACCTCCCC-3′
	GH29[b]		5′-**biotin**-GAGCTGCAGGTAGTTGTGTCTGCACAC-3′
	GH28[c]		5′-**FITC**-CTCGGATCCGCATGTGCTACTTCACCAA-3′α

[a] Primers used in the outer multiplex PCR.
[b] Primers used in the inner region-specific PCR.
[c] Primers for sequencing on the immobilized strand.
[d] Primers for sequencing on the supernatant strand.
[e] The PCR primers RIT596-599, RIT474, DB130, DB131, GH28, GH29, and BECE 3-5 are identical to previously published oligonucleotides primers (21–23).

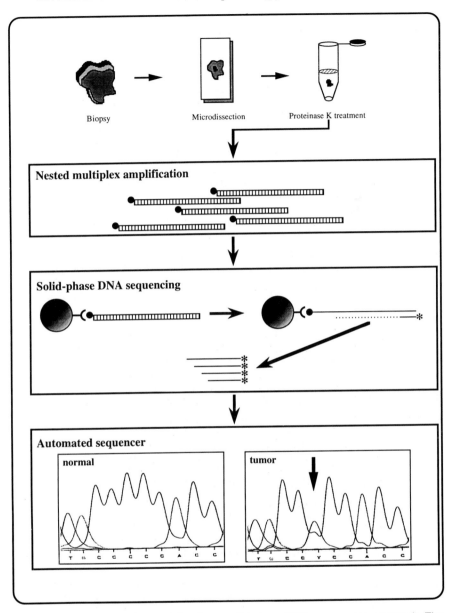

Figure 1. Schematic overview of the direct solid phase DNA sequencing approach. The microdissected sample is, after proteinase K treatment, amplified in a nested multiplex PCR. The amplicons are then analysed by solid phase DNA sequencing for analysis of tumour (mutated position) and normal tissue (wild-type sequence).

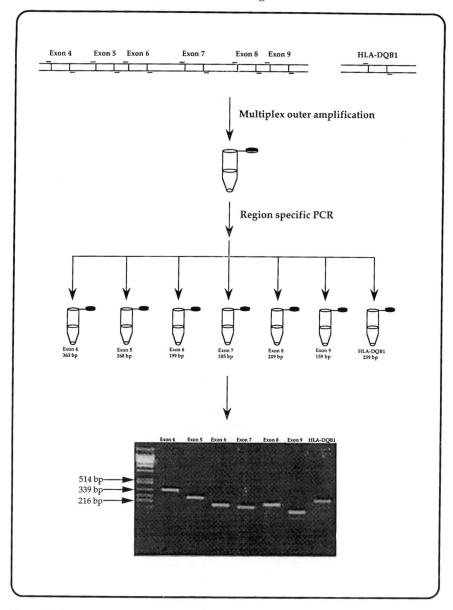

Figure 2. Schematic overview of the multiplex PCR for simultaneous amplification of exons 4–9 of the human p53 gene and the HLA-DQB1 locus. The microdissected material is amplified with seven pairs of outer primers in one tube. The generated mixtures are templates for the region-specific amplification in separate tubes. Gel electrophoresis analysis of the p53 and HLA fragments after inner amplification and ethidium bromide staining. Lane 1, DNA marker digested with *Pst*I; lane 2, exon 4 (363 bp); lane 3, exon 5 (268 bp); lane 4; exon 6 (199 bp); lane 5, exon 7 (185 bp); lane 6, exon 8 (209 bp); lane 7, exon 9 (159 bp); lane 8, HLA-DQB1 (259 bp).

4. DNA sequencing of the amplified fragments

The inner region-specific amplification generates biotinylated fragments applicable to solid phase sequencing using streptavidin-coated paramagnetic beads. After immobilization onto the solid phase, two single-strand templates for sequencing can be generated by strand-specific separation using sodium hydroxide. Thus, both the biotinylated strand, which remains bound to the solid support, and the eluted strand in the supernatant provide a suitable template for dideoxy sequencing. This approach facilitates the use of T7 DNA polymerase to obtain even peaks and therefore allows for quantitative determination of polymorphic positions. However, using standard conditions in the DNA sequencing reactions artefactual peaks can appear, making the sequence difficult to interpret. An example of primary data from sequencing

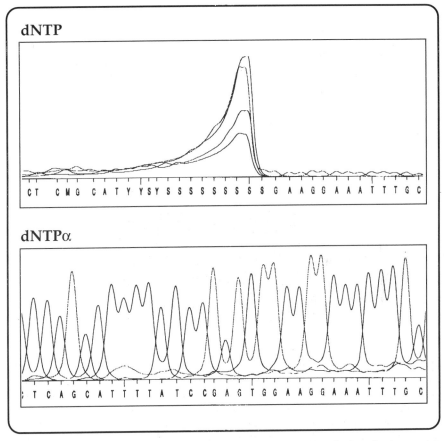

Figure 3. Sequence comparison using dNTPs and dNTPαs. The displayed primary data are generated by direct sequencing of exon 6 (p53) using standard dNTPs and dNTPαs (deoxynucleoside α-thiotriphosphates), in the sequencing reactions, respectively.

exon 6 of the p53 gene is shown in *Figure 3*. Although, it is possible to resolve this region by sequencing the complementary strand, we have replaced the common dNTPs by deoxynucleoside α-thiotriphosphates (dNTPαs) in the sequencing reactions (19). A marked improvement in resolution can be achieved upon introduction of these nucleotide analogues, which help to resolve background and stops in the p53 sequence (see *Figure 3*). Although the molecular mechanism for the improved processitivity of the polymerase through these regions is unclear, it is evident that the incorporation of dNTPαs improves the primary data substantially.

The sequencing protocol described below can be employed both for manual and automated sequencing using an automated laser fluorescent sequencing apparatus (ALF) system, but the protocol can easily be modified to other conditions. In principle the sequencing primers to be used can either be custom-designed as for p53 (complementary to a sequence inside the target DNA being amplified) or a universal primer complementary to a sequence introduced by a 'handle' sequence in one of the inner PCR primers. An example of the results of sequencing a part of the p53 gene using 5′ end fluorescent labelled sequencing primer is presented in *Figure 1*. Sequences from the automated laser fluorescent sequencing apparatus are manually edited at the polymorphic sites, and can subjectively be evaluated to determine sequence variations at polymorphic sites as earlier described by Leitner and co-workers (7).

Protocol 3. Preparation of streptavidin-coated magnetic beads

Equipment and reagents

- Dynabeads M-280 streptavidin (10 mg/ml) (Dynal AS, Oslo, Norway)
- Neodymium–iron–boron magnet (MPC Dynal AS, Oslo, Norway)
- Washing/binding solution: 10 mM Tris–HCl pH 7.5 (20°C), 1 mM EDTA, 2 M NaCl supplemented with Tween 20 (0.1% final conc.), and β-mercaptoethanol (1 mM, final conc.)

Method

1. Resuspend the beads by pipetting. Use 20 μl (200 μg) of resuspended beads per PCR template, pipette the suspended beads into a clean 1.5 ml microcentrifuge tube. The beads may be washed in bulk for the total number of PCR templates which are to be purified.

2. Place the tube in the magnetic holder and allow the beads to adhere to the magnet at the side of the tube. Remove the supernatant using a pipette (do not remove the tube from magnetic holder).

3. Add an equal volume of washing/binding solution and gently pipette to resuspend the beads.

4. Repeat step 2, using the magnetic holder, allowing the beads to adhere to the side of the tube, and remove the supernatant.

5. Resuspend the beads in washing/binding solution with twice the original volume (i.e. 40 μl). The bead concentration is now 5 μg/μl.[a]

Protocol 4.   Immobilization of PCR products

Reagents

- 1 × TE buffer: 10 mM Tris–HCl pH 7.5 (20 °C), 1 mM EDTA
- 0.05 M NaOH: use a 1.000 ± 0.005 M volumetric solution of NaOH for accurate dilution, aliquot and store at –20 °C
- 0.1667 M HCl: use a 1.000 ± 0.005 M volumetric solution of HCl for accurate dilution, aliquot and store at –20 °C

Method

1. Pipette 40 μl of the total 50 μl PCR amplification reaction into a fresh 1.5 ml microcentrifuge tube and add 40 μl of the pre-washed Dynabeads.

2. Incubate at room temperature for 15 min. Mix during the immobilization reaction once or twice by gentle pipetting or tapping.

3. Collect the beads, by moving the vials to the magnetic holder, and remove the supernatant with a pipette.

4. Wash the beads once with 50 μl washing/binding solution.

5. Wash once with 50 μl 1 × TE buffer. Remove the 1 × TE buffer carefully, avoid droplets on the walls and the bottom of the tube.

6. Resuspend the beads in *exactly* 10 μl of 0.05 M NaOH.

7. Incubate in room temperature for 5 min.

8. Collect the beads (now with only ssDNA attached) by placing the tube in the magnetic holder and transfer the 10.8 μl of NaOH supernatant (containing the non-biotinylated strand) to a clean tube. Neutralize the NaOH supernatant with 5.8 μl 0.1667 M HCl and 2 μl sterile water and mix *immediately*.

9. Resuspend the beads in 18 μl diluted washing/binding solution (dilute 20 times in water) or the appropriate buffer for the sequencing protocol (*Protocol 5*).

Protocol 5.   DNA sequencing using T7 DNA polymerase and labelled primers

Equipment and reagents

- A 6% polyacrylamide sequencing gel and electrophoresis equipment or automated sequencers (Pharmacia ALF, Uppsala, Sweden)
- Extension buffer: 300 mM citric acid pH 7.0 (20 °C), 318 mM DTT, 40 mM MnCl$_2$
- Stop solution: shake 100 ml formamide with 5 g Amberlite MB-1 resin and 300 mg

Protocol 5. *Continued*

dextran blue for 30 min; filter through 0.45 μm pore-size filter
• Sequencing primers specific for the p53 and HLA regions[a]
• Annealing buffer for single-labelled sequencing primer: 280 mM Tris–HCl pH 7.5 (20°C), 100 mM MgCl₂—the label could be either ³²P for radioactive sequencing or one dye fluorescent label such as fluorescein isothiocyanate, FITC (Pharmacia, Uppsala, Sweden)

• T7 DNA polymerase with enzyme dilution buffer (Pharmacia, Uppsala, Sweden)
• Four nucleotide mixes each containing: 40 mM Tris-HCl pH 7.5 (20°C), 50 mM NaCl, 1 mM of each 2' deoxynucleoside 5'-O-(1-thiotriphosphate), (dNTPα) (Pharmacia Biotechnology, Uppsala, Sweden), and 5 μM of one specific ddNTP—thus the 'A' mix contains 5 μM ddATP, the 'C' mix contains 5 μM ddCTP, the 'G' mix contains 5 μM ddGTP, and the 'T' mix contains 5 μM ddTTP

Method

1. Add diluted beads with immobilized single-strand DNA or eluted single-strand DNA to a fresh tube.

2. Add 2 μl (1 pmol) labelled primer.

3. Add 2 μl of annealing buffer and mix gently with a pipette. Incubate at 65°C for 10 min. Mix gently and leave to cool at room temperature for at least 10 min, mix two or three times during cooling.

4. Add 5 μl of extension buffer and mix gently.

5. Dilute the T7 DNA polymerase to 1.5 U/μl using *cold* dilution buffer; 2 μl of this diluted stock solution will be required for each template. Keep the diluted stock solution (1.5 U/μl) on ice.

6. Label four new tubes 'A', 'C', 'G', and 'T'. Dispense 2.5 μl of the corresponding dNTP/ddNTP sequencing mixes into the tubes.

7. Add 2 μl of the T7 DNA polymerase diluted stock solution (from step 5) to the template mixture (from step 4) and mix gently. Immediately add 4.5 μl of this mixture to each of the nucleotide sequencing mixes.

8. Incubate at 39°C for 5 min.

9. Add 8 μl of stop solution to each reaction and mix gently.

10. Incubate at 95°C for 5 min and then put the tubes on ice.

11. Load the samples onto 6% polyacrylamide sequencing gel.[b]

[a] Aliquots of labelled primers should be stored at –20°C (avoid repeated freeze–thawing).
[b] The sequencing reactions may be stored at –20°C if not loaded immediately. Prior to loading, heat the samples to 95°C for 5 min and place on ice.

5. Discussion

The manual interactions can in the future be further minimized by integration of the 'unit operations' (i.e. PCR set-up, PCR, PCR product quality control,

DNA sequencing reactions) onto a robotic working table. This together with new reagents will improve the quality and robustness of the DNA sequencing technology and therefore be well suited for routine clinical applications.

Acknowledgements

Supported by the Göran Gustafsson Foundation for Research in Natural Sciences and Medicine.

References

1. Lowe, S., Bodis, S., McClatchey, A., Remington, L., Ruley, H. E., Fisher, D. E., *et al.* (1994). *Science*, **266**, 807.
2. Harvey, M., Vogel, H., Morris, D., Bradley, A., Bernstein, A., and Donehover, L. A. (1995). *Nature Genet.*, **9**, 305.
3. Grompe, M. (1993). *Nature Genet.*, **5**, 111.
4. Sanger, F., Nicklen, S., and Coulson, A. R. (1977). *Proc. Natl. Acad. Sci. USA*, **74**, 5463.
5. Wahlberg, J., Albert, J., Lundeberg, J., Fenyö, E-M., and Uhlén, M. (1991). *AIDS Res. Hum. Retrov.*, **7**, 983.
6. Wahlberg, J., Albert, J., Lundeberg, J., Cox, S., Wahren, B., and Uhlén, M. (1992). *FASEB J.*, **6**, 2843.
7. Leitner, T., Halapi, E., Scarletti, G., Rossi, P., Albert, J., Fenyö, E-M., *et al.* (1993). *BioTechniques*, **15**, 120.
8. Larder, B. A., Kohli, A., Kellam, P., Kemp, S. D., Kronick, M., and Henfrey, R. D. (1993). *Nature*, **365**, 671.
9. Hultman, T., Bergh, S., Moks, T., and Uhlén, M. (1991). *BioTechniques*, **10**, 84.
10. Wahlberg, J., Holmberg, A., Bergh, S., Hultman, T., and Uhlén, M. (1992). *Electrophoresis*, **13**, 547.
11. Holmberg, A., Fry, G., and Uhlén, M. (1993). In *Automated DNA sequencing and analysis techniques* (ed. C. Venter), p. 139. Academic Press, London.
12. Hultman, T., Ståhl, S., Hornes, E., and Uhlén, M. (1989). *Nucleic Acids Res.*, **17**, 4937.
13. Hedrum, A., Pontén, F., Ren, Z., Lundeberg, J., Pontén, J., and Uhlén, M. (1994). *BioTechniques*, **17**, 1.
14. Ren, Z-P., Hedrum, A., Pontén, F., Nistér, M., Ahmadian, A., Lundeberg, J., *et al.* (1996). *Oncogene*, **12**, 765.
15. Westberg, J., Nordin, G., Truedsson, L., Sjöholm, A., and Uhlén, M. (1995). *Genomics*, **29**, 1.
16. Pettersson, B., Johansson, K-E., and Uhlén, M. (1994). *Appl. Env. Microbiol.*, **60**, 2456.
17. Odeberg, J., Yun, Z., Sönnerborg, A., Uhlén, M., and Lundeberg, J. (1995). *J. Clin. Microbiol.*, **33**, 1870.
18. Yun, Z., Odeberg, J., Lundeberg, J., Weiland, O., Uhlén, M., and Sönnerborg, A. (1996). *J. Infect. Dis.*, **173**, 992.
19. Berg, C., Hedrum, A., Holmberg, A., Pontén, F., Uhlén, M., and Lundeberg, J. (1995). *Clin.Chem.*, **41**, 1461.

20. Hollstein, M., Sidransky, D., Vogelstein, B., and Harris, C. C. (1991). *Science*, **253**, 49.
21. Kovach, J. S., McGovern, R. M., Cassady, J. D., Swanson, S. K., Wold, L. E., Vogelstein, B., *et al.* (1991). *J. Natl. Cancer Inst.*, **83**, 1004.
22. Bugawan, T. L. and Erlich, H. A. (1991). *Immunogenetics*, **33**, 163.
23. Lehman, T. A., Bennet, W. P., Metcalf, R. A., Welsh, J. A., Ecker, J., Modali, R. V., *et al.* (1991). *Cancer Res.*, **51**, 4090.

List of suppliers

Amersham

Amersham International plc., Lincoln Place, Green End, Aylesbury, Buckinghamshire HP20 2TP, UK.

Amersham Corporation, 2636 South Clearbrook Drive, Arlington Heights, IL 60005, USA.

Amicon, Amicon Inc., 72 Cherry Hill Dr., Beverly, MA 01915, USA.

Anderman

Anderman and Co. Ltd., 145 London Road, Kingston-Upon-Thames, Surrey KT17 7NH, UK.

ATTO, ATTO Corp., 2–3, Hongo 7-chome, Bunkyo-ku, Tokyo 113, Japan.

Avitech Diagnostics (Formerly AT Biocem, ATGC), 30 Spring Mill Drive, Malvern, PA 19355, USA.

Beckman Instruments

Beckman Instruments UK Ltd., Oakley Court, Kingsmead Business Park, London Road, High Wycombe, Bucks HP11 1J4, UK.

Beckman Instruments Inc., PO Box 3100, 2500 Harbor Boulevard, Fullerton, CA 92634, USA.

Becton Dickinson

Becton Dickinson and Co., Between Towns Road, Cowley, Oxford OX4 3LY, UK.

Becton Dickinson and Co., 2 Bridgewater Lane, Lincoln Park, NJ 07035, USA.

Bio

Bio 101 Inc., c/o Statech Scientific Ltd, 61–63 Dudley Street, Luton, Bedfordshire LU2 0HP, UK.

Bio 101 Inc., PO Box 2284, La Jolla, CA 92038–2284, USA.

Bio-Rad Laboratories

Bio-Rad Laboratories Ltd., Bio-Rad House, Maylands Avenue, Hemel Hempstead HP2 7TD, UK.

Bio-Rad Laboratories, Division Headquarters, 3300 Regatta Boulevard, Richmond, CA 94804, USA.

Boehringer Mannheim

Boehringer Mannheim UK (Diagnostics and Biochemicals) Ltd, Bell Lane, Lewes, East Sussex BN17 1LG, UK.

Boehringer Mannheim Corporation, Biochemical Products, 9115 Hague Road, P.O. Box 504 Indianapolis, IN 46250–0414, USA.

Boehringer Mannheim Biochemica, GmbH, Sandhofer Str. 116, Postfach 310120 D-6800 Ma 31, Germany.

British Drug Houses (BDH) Ltd, Poole, Dorset, UK.

Clontech, CLONTECH Laboratories Inc., 1020 E. Meadow Cir., Palo Alto, CA 94303–4230, USA.

Difco Laboratories

Difco Laboratories Ltd., P.O. Box 14B, Central Avenue, West Molesey, Surrey KT8 2SE, UK.

Difco Laboratories, P.O. Box 331058, Detroit, MI 48232–7058, USA.

Du Pont

Dupont (UK) Ltd., Industrial Products Division, Wedgwood Way, Stevenage, Herts, SG1 4Q, UK.

Du Pont Co. (Biotechnology Systems Division), P.O. Box 80024, Wilmington, DE 19880–002, USA.

Dynal AS, P.O. Box 158, Skøyen, N-0212, Oslo, Norway.

European Collection of Animal Cell Culture, Division of Biologics, PHLS Centre for Applied Microbiology and Research, Porton Down, Salisbury, Wilts SP4 0JG, UK.

Falcon (Falcon is a registered trademark of Becton Dickinson and Co.).

Fisher Scientific Co., 711 Forbest Avenue, Pittsburgh, PA 15219–4785, USA.

Flow Laboratories, Woodcock Hill, Harefield Road, Rickmansworth, Herts. WD3 1PQ, UK.

Fluka

Fluka-Chemie AG, CH-9470, Buchs, Switzerland.

Fluka Chemicals Ltd., The Old Brickyard, New Road, Gillingham, Dorset SP8 4JL, UK.

FMC, FMC Bioproducts, 191 Thomaston St., Rockland, ME 04841, USA.

Gibco BRL

Gibco BRL (Life Technologies Ltd.), Trident House, Renfrew Road, Paisley PA3 4EF, UK.

Gibco BRL (Life Technologies Inc.), 3175 Staler Road, Grand Island, NY 14072–0068, USA.

Arnold R. Horwell, 73 Maygrove Road, West Hampstead, London NW6 2BP, UK.

Hybaid

Hybaid Ltd., 111–113 Waldegrave Road, Teddington, Middlesex TW11 8LL, UK.

Hybaid, National Labnet Corporation, P.O. Box 841, Woodbridge, NJ. 07095, USA.

HyClone Laboratories 1725 South HyClone Road, Logan, UT 84321, USA.

International Biotechnologies Inc., 25 Science Park, New Haven, Connecticut 06535, USA.

Invitrogen Corporation
Invitrogen Corporation 3985 B Sorrenton Valley Building, San Diego, CA. 92121, USA.
Invitrogen Corporation c/o British Biotechnology Products Ltd., 4–10 The Quadrant, Barton Lane, Abingdon, OX14 3YS, UK.
Kodak: Eastman Fine Chemicals 343 State Street, Rochester, NY, USA.
Life Technologies Inc., 8451 Helgerman Court, Gaithersburg, MN 20877, USA.
Merck
Merck Industries Inc., 5 Skyline Drive, Nawthorne, NY 10532, USA.
Merck, Frankfurter Strasse, 250, Postfach 4119, D-64293, Germany.
Millipore
Millipore (UK) Ltd., The Boulevard, Blackmoor Lane, Watford, Herts WD1 8YW, UK.
Millipore Corp./Biosearch, P.O. Box 255, 80 Ashby Road, Bedford, MA 01730, USA.
New England Biolabs (NBL)
New England Biolabs (NBL), 32 Tozer Road, Beverley, MA 01915–5510, USA.
New England Biolabs (NBL), c/o CP Labs Ltd., P.O. Box 22, Bishops Stortford, Herts CM23 3DH, UK.
Nikon Corporation, Fuji Building, 2–3 Marunouchi 3-chome, Chiyoda-ku, Tokyo, Japan.
Perkin-Elmer
Perkin-Elmer Ltd., Maxwell Road, Beaconsfield, Bucks. HP9 1QA, UK.
Perkin-Elmer Ltd., Post Office Lane, Beaconsfield, Bucks, HP9 1QA, UK.
Perkin-Elmer-Cetus (The Perkin-Elmer Corporation), 761 Main Avenue, Norwalk, CT 0689, USA.
Pharmacia Biotech Europe Procordia EuroCentre, Rue de la Fuse-e 62, B-1130 Brussels, Belgium.
Pharmacia Biosystems
Pharmacia Biosystems Ltd. (Biotechnology Division), Davy Avenue, Knowlhill, Milton Keynes MK5 8PH, UK.
Pharmacia LKB Biotechnology AB, Björngatan 30, S-75182 Uppsala, Sweden.
Promega
Promega Ltd., Delta House, Enterprise Road, Chilworth Research Centre, Southampton, UK.
Promega Corporation, 2800 Woods Hollow Road, Madison, WI 53711–5399, USA.
Qiagen
Qiagen Inc., c/o Hybaid, 111–113 Waldegrave Road, Teddington, Middlesex, TW11 8LL, UK.
Qiagen Inc., 9259 Eton Avenue, Chatsworth, CA 91311, USA.
Schleicher and Schuell
Schleicher and Schuell Inc., Keene, NH 03431A, USA.

Schleicher and Schuell Inc., D-3354 Dassel, Germany. Schleicher and Schuell Inc., c/o Andermann and Company Ltd.

Shandon Scientific Ltd., Chadwick Road, Astmoor, Runcorn, Cheshire WA7 1PR, UK.

Sigma Chemical Company

Sigma Chemical Company (UK), Fancy Road, Poole, Dorset BH17 7NH, UK.

Sigma Chemical Company, 3050 Spruce Street, P.O. Box 14508, St. Louis, MO 63178–9916.

Sorvall DuPont Company, Biotechnology Division, P.O. Box 80022, Wilmington, DE 19880–0022, USA.

Stratagene

Stratagene Ltd., Unit 140, Cambridge Innovation Centre, Milton Road, Cambridge CB4 4FG, UK.

Strategene Inc., 11011 North Torrey Pines Road, La Jolla, CA 92037, USA.

United States Biochemical, P.O. Box 22400, Cleveland, OH 44122, USA.

Wellcome Reagents, Langley Court, Beckenham, Kent BR3 3BS, UK.

Index